ENVIRONMENTAL HEALTH – PHYSICAL, CHEMICAL AND BIOLOGICAL FACTORS SERIES

ENVIRONMENTAL REGULATION: EVALUATION, COMPLIANCE AND ECONOMIC IMPACT

For permission to use material from this book please contact us:
Telephone 631-231-7269; Fax 631-231-8175
Web Site: http://www.novapublishers.com

LIBRARY OF CONGRESS CATALOGING-IN-PUBLICATION DATA

Environmental regulation : evaluation, compliance and economic impact / editors, Diederik Meijer and Fillipus De Jong.
p. cm.
Includes bibliographical references and index.
ISBN 978-1-60741-645-6 (hardcover : alk. paper)
1. Environmental policy--Cross-cultural studies. 2. Economic development--Environmental aspects--Cross-cultural studies. I. Meijer, Diederik J. W. II. Jong, Fillipus de.
HC79.E5.E584 2009
333.7--dc22

2009030968

Published by Nova Science Publishers, Inc. ✢ New York

CONTENTS

PREFACE

During the last few decades, development of the environmental movement has been placing more and more pressure on firms to acknowledge, characterize, analyze and report upon environmental issues and impacts. Many organizations and enterprises are aware that environmental issues are becoming urgent in all aspects of social life. In an effort to protect the environment, environmental management measures have been introduced on both a mandatory and voluntary basis. Perceiving the benefits of compliance is crucial for firms to act in accordance with environmental laws and regulations. This new book gathers the latest research from around the globe in this field of study and focuses on related topics such as incentive-compatible targeting to improve effectiveness of environmental policies, antigenic/allergenic rubber proteins and environmental regulations, Chernobyl consequences and the efforts to reduce greenhouse gas emissions.

Chapter 1 - After the Chernobyl accident, along with serious studies, a great number of publications appeared overestimating its medical consequences. The following methodological flaws can be found in such studies: interpretation of spontaneous diseases such as radiation-induced; indication of radioactivity or dose levels without consideration of the natural background radiation; and conclusions about morbidity increase without statistically correct comparison with other regions. In some studies, high figures were obviously caused by non-random material selection or inaccurate morphological assessment. Some studies are based on small samples or singular observations and, providing no significant information, create an exaggerated impression about the consequences of the accident. Overt misquoting and misinterpretation of statistical data can be found in some publications. Several examples follow.

Chapter 2 - With more and more attention given to the study of microorganisms in oil fields in recent years, a variety of molecular biology and biotechnology techniques are also applied. This chapter describes the most commonly-used molecular biology techniques and related biotechnological advances in oil fields research at present, which are based mainly on 16s rRNA, such as 16S rRNA sequencing, Denaturant Gradient Gel Electrophoresis (DGGE), Terminal Restriction Fragment Length Polymorphism (T-RFLP) and so on. The chapter also points out their advantages and disadvantages, and introduces some of the microorganisms of greatest concern in research.

Chapter 3 - Targeting is advocated as one of the main strategies to improve effectiveness of environmental policies. Among others, this is very relevant for EU agri-environmental schemes (AESs) through which payments are provided to farmers for the provision of environmental goods. Targeting requires support by way of appropriate knowledge systems and zoning, as well as the definition of priorities for funding when subsidies are involved. However, targeting often fails due to a lack of consideration of the economic incentives to participate in target areas. This is partly motivated by the fact that agents have private information on their compliance costs which is undisclosed to the regulator. The objective of this paper is to design a model of incentive compatible targeting strategies under asymmetric information. The model is then used to evaluate the impact of targeting strategies on designing agri-environmental contracts under adverse selection in different agricultural policy scenarios. The results suggest that targeting choices directly affect the optimal way contracts are designed and that this should be considered in the policy design process in order to avoid a failure of the targeting mechanisms. In the case analysed, different agricultural policy scenarios can significantly affect AES policy design.

Chapter 4 - Austria as a member state of the European Union (EU) committed itself to a reduction of Greenhouse Gas Emissions (GHG) by 13% (based on 1990 emissions) until the Kyoto protocol observation period of 2008-2012 while the European Union has the target to reduce GHG emissions overall by 8%. Austria as a small open economy may face several problems in achieving an ambitious reduction goal. However, there emerge a number of especially interesting advantages of pursuing strict environmental policies. The paper describes the current state of climate policies in Austria and gives an overview of environmental regulations both at the national and EU level. Austria has implemented the EU regulatory framework, e.g. by taking part in the emission allowance trading scheme, and by implementing national policies. However, efforts have not been sufficient since Austria did not fulfil the goals of reducing GHG emissions. Emissions increased by roughly 20% since 1990 instead of being reduced by 13%. The "gap" between actual and Kyoto emissions would have to be filled by strict environmental policies. To the contrary, Austrian politics does not place an emphasis on climate change policies but plans to spend over EUR 400 m on purchasing "hot air" at the international carbon markets in order to offset the increase in Austrian emissions until 2012. The paper discusses other policy options and their economic impacts. An overview shows that – besides purchasing emission allowances on international markets – domestic policies would lead to higher economic benefits. For instance, policy options include increasing energy efficiency and using more renewable energy sources. Special emphasis is laid on the specifics of strict environmental policies in small open economies as new environmental technologies – with Austria currently being a net-importer of such technologies – could increase the international competitiveness of Austrian energy technologies on the international markets.

Chapter 5 - The current study seeks to assess the perceived benefits of compliance with environmental regulations by multinational corporations (MNCs) operating in Vietnam. Scott's (2001) principle of "three pillars of institutions" is used to synthesize literature on compliance into a conceptual framework of the benefits of compliance with environmental laws and regulations. The study was based on face-to-face interviews with environmental managers of MNCs operating in Vietnam. The mean importance rating, t-test of the mean and factor analysis were used to test the validity of the benefits identified in the literature. It was found that firms had derived benefits from environmental compliance which were driven by cost-benefit calculations, moral values and social pressures.

Chapter 6 - Ultraviolet (UV) irradiation represents a significant environmental and occupational hazard that can cause acute and chronic inflammatory changes in the exposed skin and cornea. The inflammatory changes of acute exposure include erythema (sunburn) of the skin and photokeratitis of the cornea. Chronic exposure to solar UV irradiation leads to photoaging, immunosuppression and ultimately carcinogenesis in the skin. After skin and cornea damage by UV radiation, these tissues are known to secrete a number of cytokines, including interleukin (IL)-1, IL-6 and tumor necrosis factor (TNF)-α. Macrophage migration inhibitory factor (MIF) was originally identified as a lymphokine that concentrates macrophages at inflammatory loci, and it is a potent activator of macrophages *in vivo* which is considered to play an important role in cell-mediated immunity. Since the molecular cloning of MIF cDNA, MIF has been re-evaluated as a proinflammatory cytokine and pituitary derived hormone that potentiates endotoxemia. MIF is ubiquitously expressed in various tissues, including the skin and cornea. This article reviews the latest findings on the roles of MIF with regard to UV-induced damage in the skin and cornea.

Chapter 7 - Recent researches in central-southern Italy demonstrated that carbonate aquifers may behave as compartmentalized systems, due to the barrier action of some low-permeability fault zones that partially impede groundwater flow. This conceptual model is different from models generally applied in karst aquifers where faults act as drains, due to their enhanced hydraulic conductivity.

Due to this discontinuous heterogeneity, these aquifers behave as basin-in-series systems where seasonal springs occur along some fault zones. Thus, the groundwater generally flows towards a main perennial spring and a number of seasonal springs. Moreover, in some basin-in-series systems relationships between adjacent compartments may change during each hydrologic year, due to differences in head fluctuation over time. Hence, the basin that feeds a spring may behave as a non-static system, and its boundaries in high flow may be different from boundaries in low flow.

To date, no specific strategies have been defined by Italian Law to protect perennial and seasonal springs against pollution in such carbonate environments. The present chapter will focus the attention on the correct methodology to update and refine existing protection measures, with emphasis on microbiological contamination.

Chapter 8 - Health and safety issues associated with the antigenic/allergenic proteins inherent in natural latex continue to affect Government policy and practice. Although many of the issues associated with Hevea Brasiliensis natural rubber latex are well documented and understood, industrial society's magnificent exploitation and comfortable dependence on such a material continues to stress human health. Agencies including the Food and Drug Administration, the Environmental Protection Agency, and the Centers for Disease Control continue to develop and implement regulations, policies, and procedures for Hevea Brasiliensis natural rubber latex. Government agencies who have been emboldened to guide health and safety have shown contradictions in policy with regard to Hevea-Brasiliensis natural rubber latex. In the prevention of chronic disease and mental health disorders, we should question if such agencies should be in the business of the proliferation of natural rubber latex, without regard for the antigenic proteins therein. Finally, the increased incidence of allergic disease in industrialized societies presents the ultimate challenge for the environmental regulation of Hevea Brasiliensis natural rubber latex.

Chapter 9 - *Artemisia annua* is currently sole herbaceous biomass for industrial manufacture of artemisinin, an antimalarial sesquiterpene lactone with the unique endoperoxide architecture. Due to presence in trace amount, artemisinin has been targeted for *in planta* overproduction by genetic modification of *A. annua*. Beneficial from such pursuits, the overall enzymatic cascades involving artemisinin biogenesis have been elucidated and a dozen of critical artemisinin responsible genes identified. Consequently, transgenic *A. annua* plants with enhanced artemisinin production are available although substantial and profound potentials in artemisinin accumulation expected. Alternatively, due to conservation of the entire terpene pathways among higher plants and eukaryotic or even prokaryotic microbes, re-establishment of extended or diverted pathways toward *de novo* microbial artemisinin production has been eagerly attempted in genetically tractable microbes. In such aspect, a suit of downstream pathway genes specific for artemisinin biogenesis have been transplanted from *A. annua* into *Sacchromyce cerevisiae* and *Escherichia coli*, in which a series of incredible amounts of artemisinin precursors manufactured. The next-step goal is to further accelerate forward the total artemisinin biosynthesis through biotransformation of the artemisinin precursor(s) either *in vivo* or *in vitro*. For this purpose, the putative oxidant sink molecule capable of quenching the reactive oxygen species (ROS), in particular, the singlet oxygen ($^1O_2$), must be produced, in a large scale, in genetically modified microbes or transgenic *A. annua* plants. Whether dihydroartemisinic acid or artemisinic acid is such a $^1O_2$-scavenging direct intermediate has not been convinced, but conversion from dihydroartemisinic acid or artemisinic acid to artemisinin recognized as a bottleneck for artemisinin biosynthesis and versatile strategies aiming at breaking the rate-limited step enthusiastically pursued in *A. annua*, for example, by utilization of the primary abiotic or biotic stress signals or secondary stress signal transducers. These achievements should benefit our future intervention with the homeostatic tempo-spatial regulation mode of genetic background-based and environment-dependent artemisinin accumulations. This article introduces, from the genomics, transcriptomics, proteomics and metabolomics, the updated literatures describing the relationship between artemisinin biosynthetic gene overexpression and subsequent artemisinin overproduction as well. It should shed light on further elucidation of the intrinsic rule and mechanism underlying that artemisinin biochemical synthesis is fine-tuned by the genetic and environmental regulators, and should also urge the researchers all over the world more intensively investigating the intriguing *A. annua* plant that has implications in the medicinal and aromatic industries.

Chapter 10 - The early stages of environmental policy implementation were characterized by the regulation of point source pollution. This is partly explained by the easy identification of these sources and the broad and strong political support for their regulation. In contrast, the regulation of non-point sources began much later and is still being implemented. For instance, international agreements to reduce water and gaseous emissions within the EU are likely to lead to more regulations in European countries in coming years. The European Water Framework Directive (WFD) has great power to reduce non-point pollution in European member states. This initiative is supported by European Environmental Agency findings (2006) that point to agricultural non-point pollution as the primary cause of water quality deterioration in many European watersheds. From an economic perspective, agricultural non-point control involves difficult planning problems characterized by complexity, uncertainty and policy conflicts. The major feature of non-point pollution is that emissions are either not observable or cannot be observed at a reasonable cost. Therefore it is impossible to attribute emissions to particular polluters, and the use of first-best instruments is infeasible. Unfortunately, the economic literature does not clearly indicate which are the optimal second-best instruments to regulate non-point sources.

In: Environmental Regulation: Evaluation, Compliance . . . ISBN: 978-1-60741-645-6
Ed: Diederik Meijer and Fillipus De Jong

Chapter 1

OVERESTIMATION OF CHERNOBYL CONSEQUENCES: MOTIVES AND MECHANISMS

Sergej Jarqin
People's Friendship University of Russia

INTRODUCTION

After the Chernobyl accident, along with serious studies, a great number of publications appeared overestimating its medical consequences. The following methodological flaws can be found in such studies: interpretation of spontaneous diseases such as radiation-induced; indication of radioactivity or dose levels without consideration of the natural background radiation; and conclusions about morbidity increase without statistically correct comparison with other regions. In some studies, high figures were obviously caused by non-random material selection or inaccurate morphological assessment. Some studies are based on small samples or singular observations and, providing no significant information, create an exaggerated impression about the consequences of the accident. Overt misquoting and misinterpretation of statistical data can be found in some publications. Several examples follow.

EXAMPLES OF OVERESTIMATION

In an autopsy report of a Chernobyl recovery operation worker with lung carcinoma [1], electron-dense inclusions in alveolar macrophages are described without any evidence of causative relationship between these inclusions, which are probably dust particles, and radiation or carcinogenesis. Radioactivity of lung tissue, established by gamma-spectrometry, is reported to be 0.1–0.18 Bk, but the sample weight is not given, which makes this figure senseless. Nonetheless, the case is discussed as radiation-induced malignancy.

Another example is an international study, based on 15 accidental cases of lung carcinoma in patients "who had worked in nuclear industry or resided in radiocontaminated areas after the Chernobyl accident" [2]. Morphological and molecular-genetic features of radiation-induced carcinoma are discussed and a conclusion about its "poor prognosis" is made on the basis of the non-representative sample. Radiation doses are unknown. A quotation from this article is an example of unfounded speculation (verbatim from Russian): "In case of peripheral carcinoma, the matter could concern aerosol state of carcinogenic, including radioactive, particles (Cheliabink-95; Tula region)" [2].

In a series of works by Derizhanova et al., presented at the 23rd Congress of the International Academy of Pathology in Nagoya [3] and other international meetings, renal and pulmonary changes in Chernobyl recovery operation workers (so-called liquidators) are described in autopsy material: chronic atrophic bronchitis, sclerosis of alveolar septa, emphysema, dust accumulation in macrophages etc.; then follows an unfounded conclusion that the pulmonary lesions were caused by incorporation of "radioactive Chernobyl dust." Anatomic pathologists know that unspecific changes of that kind are often seen at autopsy. Furthermore, "sclerosing glomerulonephrosis" is described in the kidneys without commentary about its extent or relation to the cause of death. Sclerosis of some glomeruli is a usual finding in arterial hypertension and atherosclerosis, without significant decrease of renal function as a rule; scattered obsolete glomeruli are often seen in kidneys taken at autopsy. Derizhanova et al. designated these changes as "pulmonary renal syndrome," a synonym of Goodpasture's syndrome. An impression of significant renal and pulmonary abnormality induced by radiation is created in this way. Another research group at the same congress, having presented no statistical data, declared recovery operation workers and residents of contaminated areas to be a risk group for stomach carcinoma and MALT-lymphoma [4].

An example of misinterpretation of statistics is provided in the article by Okeanov et al. [5] based on the data from Belarusian National Cancer Registry. The following statement is made: "A significant increase in the incidence of cancer morbidity of colon, lung, urinary bladder and thyroid gland, as well as cancers of all sites, was observed in the population of contaminated areas." The radiocontamination was maximal in Gomel and minimal in the Vitebsk regions of Belarus; the latter was used as control. An extract from Table 1 from the article [5], referring to the above-named regions, the capital city of Minsk and the whole country, is given as example. The title of the table below is reproduced from the original.

Table 1. Average incidence/year for all types of cancer (in 100,000 inhabitants), based on standardized indices, world standard

Region	Average incidence and standard error	
	1976–1985	1990–2000
Vitebsk	158.2 ± 3.24	217.9 ± 3.5
Gomel	147.5 ± 2.52	224.6 ± 6.3
Minsk city	223.5 ± 5.72	263.7 ± 1.76
Belarus	155.9 ± 3.80	217.9 ± 3.4

The difference between the figures from the Gomel and Vitebsk regions (1990–2000 years) given in the table is statistically insignificant. Incidence increase reported in the article can be explained by improved diagnostics in contaminated areas and in the population at risk after the Chernobyl accident. The proof of it is provided by the relatively high figures for Minsk, obviously caused by a more developed health care system in the capital city in comparison with rural areas. Tumor incidence in Minsk was significantly higher than in the most contaminated Gomel region and in the Minsk region surrounding the capital (Table 2).

Table 2. Comparison of cancer incidence in different areas of Belarus

Compared regions	Years	Average incidence and standard error	Statistical significance of the difference (P)
Vitebsk region vs. Gomel region	1990–2000	217.9 ± 3.5 vs. 224.6 ± 6.3	> 0.1
Minsk city vs. Gomel region	1990–2000	263.7 ± 1.76 vs. 224.6 ± 6.3	< 0.001
Minsk city vs. Gomel region	1976–1985	223.5 ± 5.72 vs. 147.5 ± 2.52	< 0.001
Minsk city vs. Minsk region	1990–2000	263.7 ± 1.76 vs. 216.6 ± 3.9	< 0.001
Minsk city vs. Minsk region	1976–1985	223.5 ± 5.72 vs. 145.3 ± 2.26	< 0.001

These quantitative relationships witness against the causative role of radiocontamination, which was higher in rural areas than in towns because of nuclide preservation in the soil and in forests, one of the main exposure sources being locally-produced foodstuffs [6].

Urologic Malignancy

Previously we commented [7-8] on several publications by Romanenko et al. [9-11], among other things, about inadequate use of the term "long-term low-dose exposure to ionizing radiation" and unification of the patients from radiocontaminated areas and from Kiev within one research group, thus creating a ground for speculation about radiation-induced malignancy in the large city. In their responses [7-8], the authors did not comment on these points. Estimated individual effective doses of radiation, received by Kiev inhabitants in the year following Chernobyl accident (external irradiation 3 mSv plus internal irradiation 1.1 mSv, decreasing in the following years) [12], were comparable with the global average annual dose due to natural background radiation. In contaminated areas, average effective doses received by inhabitants were around 40 mSv in the year after the accident but fell to less than 10 mSv in the following years, thus remaining under the recommended upper limit for occupational workers, which is 20 mSv/year (averaged over the period of five years, not exceeding 50 mSv in any single year) [13].

Some publications of Romanenko et al. contain provable overestimation of radiation-induced abnormality in the kidneys [9] and urinary bladder [10-11]. For example, the article [9] is based on the renal tumors from the Chernobyl area and Kiev studied as a single group. Alterations of extracellular matrix in renal cell carcinoma are reported. The main conclusion is that long-term low-dose ionizing radiation "can promote malignant tumor progression" by the following mechanism: "significantly disrupted extracellular matrix with a lack of orchestrated communication between cells and the extracellular matrix that leads to loss of cellular differentiation." It should be noted that debate about small doses would have been applicable if the radioactivity level had been counted from zero but, in fact, it is a question not of low doses, but of a minor increase in the background radiation.

On the basis of the limited material (contaminated area plus Kiev—41 patients; control—37 patients), statistically significant differences were found for four variables: immunohistochemical expression of fibronectin, laminin, β-catenin and TGF-β1 (P values are 0.05, 0.008, 0.003 and 0.01, respectively), the integral level of statistical significance being very high for a medical study. At the same time, the tumors were obviously spontaneous because no increases in overall cancer incidence or mortality have been observed, which could be attributed to ionizing radiation (apart from the increase in thyroid cancer after childhood exposure, discussed below) [6,14]. The male/female ratio equal to 27/14, approximately the same as for renal cell carcinoma in the general population, provides additional evidence in favor of spontaneous origin. Besides, comparison between the group from the contaminated zone plus Kiev and the control is performed without standardization after the tumor grade: more than 50% of cases in the first group belong to the poorly-differentiated carcinoma (G3–G4), whereas in the control there is only one G3 case, all others being G1–G2. Without representative statistics on tumor grading, these data are misleading, creating an impression that renal carcinoma in the contaminated areas and Kiev is on average less differentiated than in the overall population. Reported differences between the groups are probably caused by the difference in differentiation grade, which in its turn could have resulted from non-random case selection.

Romanenko et al. wrote in their response to our letter to the editor that activity concentration of ^{137}Cs in the urine of the patients from contaminated areas, studied by them, was on average 6.47 Bq/liter [8]. For comparison, according to the guidelines for drinking-water quality [15], the guidance level for ^{137}Cs in the drinking water is 10 Bq/liter. This activity concentration corresponds to the recommended reference dose level equal to 0.1 mSv from one year's consumption of drinking water (assumed to be 730 liters/year). Background radiation exposures vary widely, but the average is 2.4 mSv/year, with the highest local levels being up to 10 times higher without any detected increased health risks from population studies; 0.1 mSv being thus a small addition to the background radiation [15]. Hence, activity concentration of ^{137}Cs in the urine, reported by Romanenko et al., is within the limits recommended for the drinking water. Moreover, the alkali metal cesium behaves in the organism similarly to potassium, its main excretory pathway being urine [16]. Accordingly, cesium concentration in the urine must normally be higher than in drinking water because of the renal concentration. Considering the above, reported activity concentration is too low to cause an increase in bladder malignancy or radiation-induced chronic proliferative atypical cystitis, named Chernobyl cystitis by Romanenko et al., characterized by multiple areas of urothelial dysplasia (92% in random patients from contaminated areas with benign prostatic hyperplasia) and carcinoma in situ (CIS—76%) [11]. Even higher figures were reported in another publication: 100% and 86%, correspondingly [17].

The argument that patients with benign hyperplasia of prostate had "urinary retention and therefore presumably high radiation exposure to the urothelium" [8] is hardly sustainable because beta-particles (electrons) emitted as a result of ^{137}Cs radioactive decay predominantly with the energy 0.51 MeV [18] penetrate in water for only about 2 mm [19]. In fact, the average distance is shorter: tables or data pertinent to radioactive decay always list the maximum beta-particle energy. In calculations of energy delivered by beta particles to any medium, the mean energy rather than maximum is needed. Mean beta energy varies between 0.2 and 0.4 of the maximum, which in case of ^{137}Cs is 0.51 MeV (92% of beta particles) and 1.17 MeV (8%); the estimated mean value being 0.23 MeV [20]. Gamma photons, having comparable energies (0.662 MeV in case of ^{137}Cs), would cause much lesser radiation damage in the thin urothelial layer than beta particles because of greater penetration distance in tissue (half-value layer more than 6 cm) [20]. In other words, for the beta rays filling of the urinary bladder is of little importance, whereas gamma rays by comparable energies would cause much lesser damage of the thin urothelial layer. Therefore, excessive filling of the bladder is of minor significance for radiation exposure of the urothelium from ^{137}Cs contained in the urine in activity concentrations given by the authors.

The flat CIS is most often found in association with papillary or invasive carcinoma. If left untreated, about a half (or more, according to some researchers) of CIS cases will progress to invasive carcinoma [21]. Therefore, high frequency of bladder CIS, found by Romanenko et al. in random patients with benign prostatic hyperplasia, is incompatible with the bladder cancer incidence in Ukraine (50.3 cases per 100,000 inhabitants) reported by the same authors. Romanenko et al. argued that they used "the recent WHO classification for the histological diagnosis of bladder cancer. According to this classification, CIS includes CIS from the previously classification and severe dysplasia. Therefore, the incidence of CIS may be high compared to previous WHO classification data" [8]. Is should be commented that urothelial dysplasia can overlap with cytological abnormalities seen in reactive conditions. The terms dysplasia and CIS should be limited to those cases where cytological abnormalities are certainly neoplastic. Reactive atypias and atypias of unknown significance are best reported as such. This concept was reflected in the 1998 WHO/ISUP classification and ratified in the 2003 WHO classification [21]. Another thinkable cause of the reported high CIS incidence could be non-random selection of specimens. It should be concluded that publications by Romanenko et al. tend to overestimate significance of urologic malignancy and pre-malignant lesions in consequence of Chernobyl accident.

Thyroid Carcinoma

It would be logical to assume by analogy such an attitude also applies to the thyroid. Thyroid carcinoma (TC) in children and adolescents is the only type of malignancy, a significant increase of which in consequence of Chernobyl accident (CA) is regarded to be proven [6, 14, 22]. The reaction of the scientific community to reports of its drastic increase, begun four years after the CA, was skeptical: it had been assumed that radiation from ^{131}I is less carcinogenic to the thyroid than external radiation, and that a latent period for thyroid carcinoma after an exposure should be at least 10 years. There was also uncertainty about accuracy of the diagnoses [23]. High incidence and the short induction period were designated as unusual in the UNSCEAR 2000 report, where it is also stated that the number of thyroid cancers in children and adolescents exposed to radiation is considerably higher than expected on the basis of previous knowledge. It is assumed that other factors may be influencing the risk [14]. Improved diagnostics, registration and reporting were named among factors that could have contributed to the increased cancer incidence after the CA [22]. It is also noteworthy that exposures to ^{131}I from medical procedures have demonstrated no convincing evidence of an increased thyroid cancer risk [24]. This chapter is based on experience of histopathological practice in the former Soviet Union [25], visiting cytological and histopathological laboratories, and interviewing physicians in the northern regions of

Ukraine. Besides, information from Russian-language professional literature can shed more light on the issue.

The following figures can give an estimate of the incidence increase. In the Ukraine, before the CA, about 12 cases of TC were registered in children and adolescents yearly. During the five years preceding the CA (1981–85), a total of 59 cases of thyroid carcinoma were diagnosed among patients younger than 18 years. By the year 1997, the total number of thyroid carcinoma cases, registered in Ukraine in children and adolescents, was 577 [26]. In Belarus, TC incidence in some areas increased after the CA from 0.04 to 20–50 cases per 100,000 children [27]. From 1992 to 2002 in Belarus, Russia and Ukraine more than 4000 cases of thyroid cancer were diagnosed among persons who had been children and adolescents at the time of the accident [6].

At the same time, it is known that coverage by medical examinations of the population at risk after the CA was significantly improved. Ultrasonic thyroid screening was performed in population groups at risk, and large number of thyroid nodules was found. Equipment of histopathological laboratories was poor and outdated; excessive thickness of histological sections hindered reliable assessment of diagnostic criteria. Gross dissection of surgical specimens was often made with blunt autopsy knives, without rinsing instruments and cutting board with water, which can result in tissue deformation, contamination of the cut surface by cells and tissue fragments and other artifacts, hardly distinguishable from malignancy criteria. It could have contributed to unusually high frequency of tumor cell finding in blood vessel lumina (45%) reported in post-Chernobyl pediatric TC [27]. In many laboratories celloidin embedding was used, not allowing reliable evaluation of characteristic nuclear changes in papillary thyroid carcinoma, in particular, the ground-glass nuclei. Pathologists in Russia, having experience with thyroid tumors from radiocontaminated areas, pointed out the "low quality of histological specimens, impeding assessment of nuclei" [28].

False-positive diagnosis of TC was not excluded after cytological and histological examination. If a thyroid nodule is found during ultrasonic screening, a fine-needle aspiration biopsy (FNA) is usually performed. Thyroid FNA cytology is known to be accompanied by a certain percentage of inconclusive results (so-called grey zone): figures about 10–20 % are reported from modern clinical centers [29] but in the former Soviet Union percentage was higher, one of the causes being absence of modern literature in hospitals and laboratories. Data about sensitivity of the FNA in detecting post-Chernobyl childhood TC can be found in the dissertation of A.Iu. Abrosimov [30], a well-known specialist in this area: "In a definite or presumptive form, diagnosis of carcinoma was established in 161 from 238 cases," whereas papillary carcinoma was diagnosed correctly by FNA in 69.5% and its follicular variety was only in 36.5% of cases. As it follows from the context, presumptive diagnoses were included among correctly diagnosed cases. After receiving a cytological report in a presumptive form ("atypical cells" or "suspicion of carcinoma"), depending on the nodule size, a lobectomy or subtotal thyroid resection is performed, and the surgical specimen is sent for pathological examination. Histopathological differential diagnosis of thyroid nodules is again a problematic area. High quality of specimens, required for adequate evaluation of nuclear changes in papillary carcinoma, was not always achieved at that time. For search and evaluation of malignancy criteria of a minimally-invasive follicular carcinoma (capsular and vascular invasions) great number of sections can be needed, which have not always been made. Besides, it is known from praxis that after in toto removal of a presumed carcinoma, a pathologist can be inclined to confirming malignancy even in case of some uncertainty.

Moreover, in the 1990s some diagnostic criteria of TC were hardly known in the former Soviet Union and were not mentioned by Russian-language handbooks and monographs in use at that time [31-32]. The minimally-invasive follicular carcinoma and its diagnostic criteria were absent in Russian-language literature. One of the most significant diagnostic criteria of papillary carcinoma—ground-glass or cleared nuclei—was mistranslated as something like "watch-glass nuclei molded together" (*yadra v vide pritertykh chasovykh stekol*) and presented by the most authoritative Russian-language handbook of tumor pathology [32] as a sign not only of papillary, but also of follicular TC, for which it is not characteristic. Description of this phenomenon does not agree with international literature. Nuclear changes, characteristic of papillary carcinoma, are not visible in the illustrations of this handbook. Even less understandable comparison with a sand-glass (another mistranslation of the "ground-glass") can be encountered.

In the *Atlas of Human Tumor Pathology* [33], recently edited in Russia, the following is stated in regard to thyroid nodules: "In severe dysplasia appear cell groups with clearly visible atypia. Therefore, third grade dysplasia is considered as an obligate pre-cancer, which histologically is hardly distinguishable from carcinoma in situ". Nuclear atypia (enlargement, hyperchromatism, pleomorphism) is not regarded in modern literature as a malignancy criterion of follicular and papillary thyroid nodules, and the concepts of carcinoma in situ and dysplasia are not applied to them [34]. Cases of false-positive TC diagnosis, caused by misinterpretation of nuclear atypia as a malignancy criterion, are known. Follicular and solid varieties of papillary TC prevailed in children and adolescents after the CA [35-36]. Diagnosis of these subtypes of papillary carcinoma is largely based on the nuclear criteria, inadequate assessment of which can result in false-positive conclusions, for example, in case of well-differentiated tumors of uncertain malignant potential [37] or benign papillary nodules [38].

Physicians of histopathological and cytological laboratories in northern Ukraine mostly agreed that improved diagnostics, screening and oncological alertness contributed to high incidence figures of post-Chernobyl TC. Elevated interest to radiation-induced malignancy as a topic for research and publication, coupled with economic interests (foreign help, international scientific cooperation, etc.) was pointed out as a motive for biased presentation of data. It should be noted that foreign research partners, having scrupulously performed expensive modern tests, can unexpectedly find themselves embroiled in scientific misconduct, which has been not uncommon in Soviet science [39-40]. In particular, correlations between estimated doses and thyroid cancer incidence, reported by some authors, can be explained by uncertainties in establishing the doses and incomplete case ascertainment [16], which provided rich opportunities for manipulations with statistical data.

Remarkable observations about the post-Chernobyl attitude regarding thyroid nodules can be found in Russian-language literature: "Practically all nodular thyroid lesions in children, independently of their size, were regarded as potentially malignant neoplasms, requiring urgent surgical operation"; "Aggressiveness of surgeons contributed to the shortening of the minimal latent period" [41]. Obviously, it was not the matter of true latency shortening but of early detection. Data about verification by expert commissions of post-Chernobyl pediatric TC in Russia provided further evidence for false-positivity: "As a result of histopathological verification, diagnosis of TC was confirmed in 79.1 % of cases (federal level of verification—354 cases) and 77.9% (international level —280 cases)" [30].

Obviously, false-positive diagnoses remained undisclosed in cases not covered by verification.

Another piece of evidence in favor of false-positivity is the incidence of pediatric TC in the Bryansk region, the most radiocontaminated area in Russian Federation, having increased from zero (1986–89) up to 9 cases in 1994 and 8 in 1995, decreased back to zero in 2001 and 1 case yearly in the subsequent years 2002–2003 [41], disagreeing with prognoses that radiation-induced TC morbidity in Bryansk region inhabitants who had been children or adolescents at the time of CA would grow nearly exponentially until 2021 and beyond [42]. Analogous data were reported also from Belarus [43]. These figures are in conflict with the data obtained in survivors of atomic bombings: the most pronounced thyroid cancer risk was found among those exposed before the age of 10 years, the highest risk being observed 15–29 years after the exposure [16]. The early and "dramatic" TC increase with gradual decrease 13 years after the Chernobyl accident can be explained only by false-positivity in the early period with subsequent improvement of diagnostic accuracy. In this connection, lack of statistically significant TC increase in children born after the CA can be understood: the data pertaining to them originated from a later period, "oncological alertness" affected these children to a lesser degree, and there were no motives to artificially enhance the figures.

Research on post-Chernobyl pediatric TC is in a cul-de-sac: the data documenting its dramatic increase are felt to be inflated, but to prove it with figures would be not easy today. Arguments presented in this chapter allow concluding that the high figures were at least in part caused by improved detection of thyroid nodules with occasional false-positive diagnoses of malignancy. Besides, latent carcinomas and borderline lesions, including well-differentiated tumors of uncertain malignant potential [37] found by echo-screening and diagnosed as malignancies, could have additionally contributed to the high incidence figures. Clarification of the issue, retrospective correction of false-positive diagnoses and their preclusion in future are indicated because of the overtreatment risk. The following treatment is recommended for the children with the supposed radiation-induced TC: "radical thyroid surgery including total thyroidectomy combined with neck dissections followed by radioiodine ablation" [43]; "Thyroidectomy and lymph node dissection. Careful and complete removal of the lymph nodes is of great clinical relevance" [44].

Semipalatinsk Area

The same authors, who contributed to the overestimation of Chernobyl consequences, published similar reports also about other areas with elevated background radiation. For example, an international study [45] based on two groups of patients with lung carcinoma: first group—17 cases from the area of Semipalatinsk in Kazakhstan, where from 1949 to 1989 nuclear tests had been carried out; the second group—40 patients from areas with no elevation of radioactivity. Morbidity and mortality statistics of lung cancer in Semipalatinsk area are not given. With reference to the doses it is stated in the English summary: "17 patients (group 1) lived close to the testing area from the childhood to 1993 and were exposed to the radiation at the year dose 0.1 ber." A radiation dose unit "ber" (biological equivalent of a rad) is designated internationally as rem. The yearly individual dose of 0.1 rem is within the limits of doses obtained in many areas due to the natural background radiation. At the same time, it is known from practice that documentation in pathology departments, where tissue blocks for the study were collected, usually does not allow determining how long a person has lived in the area, let alone the doses of radiation.

Table 3. Main characteristics of material, given in article [45]

Characteristics	First Group (Semipalatinsk area)	First Group (control)
Number of cases	17	40
Carcinoma type	9 (53%)—oat-cell 2 (12%)—large-cell 4 (24%)—squamous-cell 2 (12%)—adenocarcinoma	6 (15%)—oat-cell 6 (15%)—large-cell 13 (33%)—squamous-cell 15 (38%)—adenocarcinoma
Neuroendocrine differentiation	17 (100%)	0 (0%)
Metastasizing	5 (30%)	16 (40%)

Remarkable are the figures in the fourth line, which are commented in the article: "Specific feature of the lung carcinoma in patients from the area of Semipalatinsk was the neuroendocrine differentiation of tumor cells. We have established it by means of immunohistochemical and electron-microscopic investigations." At the same time, "no neuroendocrine differentiation was shown in the control group." It means that the feature established by two independent methods appears in the first group with the frequency of 100% (17/17) and in the second group—0% (0/40). Extremely high difference between the two groups ($P<0.0001$, Fisher's exact test) confirms the supposition that the "lung cancer in persons exposed for a long time to radionuclide radiation pollution" is a particular entity, different from spontaneous lung carcinoma. Additionally, significant differences between the two groups were found also for other variables, for example, proliferation marker Ki-67.

According to the conclusion, "lung carcinoma in patients, who have resided in the area of Semipalatinsk and underwent elevated radioactivity, can be classified as neuroendocrine carcinoma." It remains unclear, how all 17 randomly selected cases could have fallen into the group of the radiation-associated neuroendocrine carcinoma. In the overall population, neuroendocrine tumors, including small cell carcinoma and carcinoids, represent only about 20–30% of all lung cancer cases [46]. Where is the conventional lung cancer, caused by smoking, industrial or other carcinogens? The distribution after age and sex in the Semipalatinsk group is typical exactly for spontaneous carcinoma: 15 from 17 patients belong to the age group of 51–70 years. Under the influence of radiation, morbidity in younger age could be expected. In particular, for spontaneous lung carcinoma is typical predominance of males because of smoking and professional carcinogens. Radioactive pollution, on the contrary, has an equal effect on both sexes. In the group from Semipalatinsk were 16 males and one female [47].

To estimate what incidence increase would have been necessary for all 17 random patients to fall with 50% probability into the group of "radiation carcinoma" we used the formula of multiplication of recurring events $P= p^n$, where P is the probability of n-times repetition of an event and p is the probability of a single event. In this case p is equal to the part of radiation cancers among all cancer cases. Substituting $P=1/2$ and $n=17$ we come to the result that all 17 patients could with 50 % probability fall into the group of radiation-induced cancer only after a 24-fold increase of morbidity in consequence of radiation. For the older population groups it would mean a real epidemic, which has never been reported. Misinterpretation or fabrication of data is thereby proven by reductio ad absurdum. The term "radiation carcinoma" and speculations about its rapid growth and "bad prognosis" [45], are used to exaggerate impression about severe medical consequences of the elevated background radiation.

Linear Non-Threshold Theory

The Linear Non-Threshold Theory (LNT) of radiation carcinogenesis provided a theoretical basis for the overestimation of Chernobyl consequences. According to the LNT, the linear dose-and-effect correlation, proven to a greater or lesser extent for higher doses, can be extrapolated down to the zero level and thus be used for predicting of radiation-induced cancer incidence. The LNT is corroborated by the following arguments: radiation has a stochastic effect; the more high-energy particles or gamma photons hit a cell nucleus, the more will be DNA damage and the higher the risk of malignant transformation [48]. This paradigm does not take into account that DNA damage and repair are persistent processes, normally being in dynamic equilibrium. Optimal for a living organism must be the radiation level, adaptation to which has occurred as a result of natural selection. This is obviously the case for other environmental factors: light and ultraviolet radiation, atmospheric pressure, etc., where deviation in either direction from the optimum is harmful. For ionizing radiation it is confirmed by experimental and epidemiological evidence in favor of the hormesis (beneficial effect of low-level exposure) [49-50], as well as lacking increase of radiation-induced abnormalities in the regions with elevated background radiation [51-52].

Natural selection is a slow process: adaptation to a changing environmental factor must lag behind its current value. Therefore, actual adaptation must correspond to some average of previous levels. It is known that natural background radiation has gradually decreased during last millions of years, main mechanisms being decay of radionuclides on the earth surface and oxygen accumulation in the atmosphere due to the photosynthesis, resulting in formation of the ozone layer, protecting against the cosmic rays. Accordingly, living organisms must be adapted to a higher background radiation than that existing today. Therefore, the LNT is not applicable to the radiation doses comparable to those received from the natural background; there can be even some 'radiation hunger'. Understanding of theses facts must contribute to a more realistic approach to anthropogenic increase of radiation and its medical implications.

Propaganda Under the Guise of Scientific Publications

Along with the articles supposedly based on research, unreliable information is spread under the guise of scientific reviewing. Examples can be found in publications by Alexey V. Yablokov [53-56]. Professor Yablokov, corresponding member of Russian Academy of Sciences, publishes his views in mass media and on the Internet [55], organizes picketing (for example, in 2008 in Geneva) with the obvious goal: to foster anti-nuclear sentiments in the public. It would be naïve to suppose that A.V. Yablokov is an independent environmentalist:

there are hardly any independent public activists in today's Russia. A politician or governmental agent guised as a scientist is a phenomenon that should be added to the known forms of scientific misconduct [57]. Misquoting and misinterpretation of statistics with the obvious goal to create exaggerated impression about Chernobyl consequences have been demonstrated in publications by A.V. Yablokov. There follow some examples of false and tendentious quotation. The photographs of the tables and text fragments, translations of which are reproduced below, as well as examples of plagiarism and manipulations with statistics, can be viewed online [58].

There follow examples of false and tendentious quotation. Figure 1 is a photograph from the book Yablokov, 2002. A translation of a fragment from Table 6 of that text is given below in Table 4.

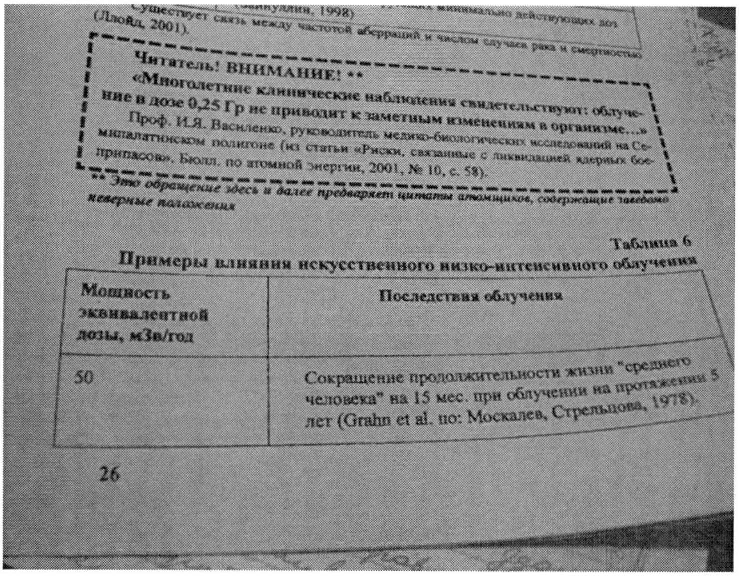

(Ллойд, 2001).

Существует связь между частотой аберраций и числом случаев рака и смертностью

Читатель! ВНИМАНИЕ! **

«Многолетние клинические наблюдения свидетельствуют: облучение в дозе 0,25 Гр не приводит к заметным изменениям в организме...»

Проф. И.Я. Василенко, руководитель медико-биологических исследований на Семипалатинском полигоне (из статьи «Риски, связанные с ликвидацией ядерных боеприпасов». Бюлл. по атомной энергии, 2001, № 10, с. 58).

*** Это обращение здесь и далее предваряет цитаты атомщиков, содержащие заведомо неверные положения*

Таблица 6

Примеры влияния искусственного низко-интенсивного облучения

Мощность эквивалентной дозы, мЗв/год	**Последствия облучения**
50	Сокращение продолжительности жизни "среднего человека" на 15 мес. при облучении на протяжении 5 лет (Grahn et al. по: Москалев, Стрельцова, 1978).

26

Figure 1. From the book by Yablokov, 2002; p. 26 [54)

Table 4. Examples of influence of low intensity artificial radiation [54]

Equivalent dose power mSv/year	Consequences of radiation
50	Reduction of life duration of an "average human" by 15 months after irradiation during five years (Grahn et al. after Moskalev, Streltsova, 1978)

In the reference list the publication by Grahn D. (1970] is quoted indirectly (Figure 2], after a Russian publication (Moskalev and Streltsova, 1987), this time with another year, which in itself is a misquoting. We have found the Russian source and translated the corresponding passage on p. 32 (Figure 3), containing a reference to the publication by Grahn D. et al., 1970: "According to the calculations by D. Grahn, G.A. Sacher, R.A. Lea et al., the radiation dose of 2,5 Sv (i.e., 0.05 Sv yearly, obtained during 50 years of work, which today is the upper limit for occupational workers), received as chronic irradiation during a very long time, can cause reduction of life duration by 15 days for an average mouse aged 100 days and by approximately 15 months for an average human aged 20 years" [59]. This passage is misquoted, and the key figure is changed (five years instead of 50), creating a wrong impression about complications of exposure to low dose radiation.

in Early Childhood Cancer. Int. J. Health Services, vol.30, #3, pp. 513 - 539.
Grahn D. 1970. Biological Effects of Protracted Low-Dose Irradiations of Mans and Animals. In R.J.M. Fry, D. Grahn, M.L. Frein,. J.H. Rust (Eds.) «Late Effect of Radiation». L., «Taylor and Francis», pp. 101-138 (цит. по: Москалев, Стрельцов, 1987).
Green P. 1990. Low-level radiation: Questions and answers (цит. по: Aubrey C., Grunberg B., Hildyard N. 1990. (Eds.) Nuclear Power: Shut it down: An information pack on nuclear power and the

Figure 2. From the book by Yablokov, 2002; from the reference list [54)

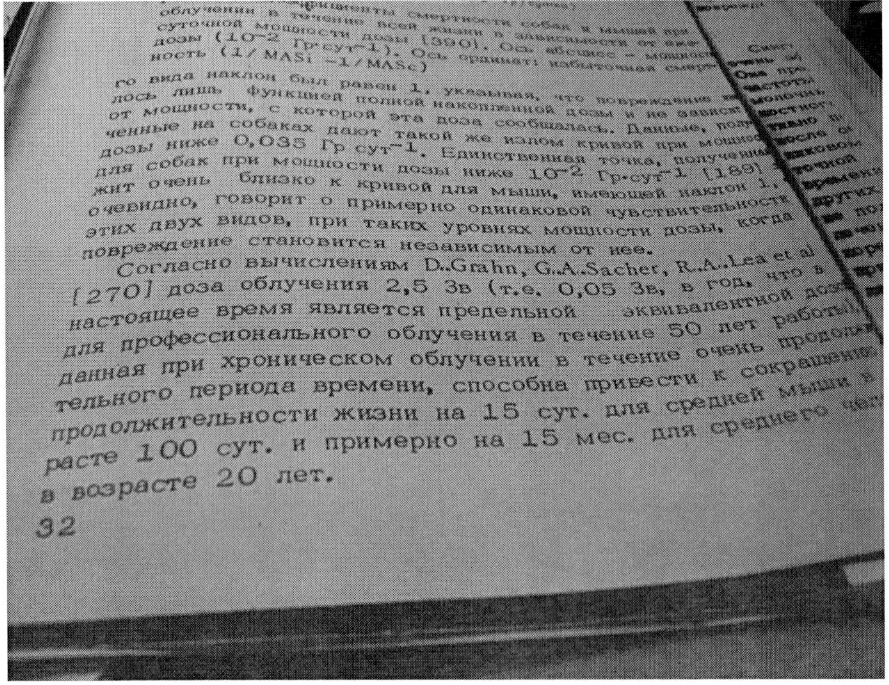
облучении в течение всей жизни в зависимости от еже-
суточной мощности дозы (390). Ось абсцисс – мощность
дозы (10^{-2} Гр·сут$^{-1}$). Ось ординат: избыточная смерт-
ность (1/MASi –1/MASc)

го вида наклон был равен 1, указывая, что повреждение я-
лось лишь функцией полной накопленной дозы и не зависит
от мощности, с которой эта доза сообщалась. Данные, полу-
ченные на собаках дают такой же излом кривой при мощнос-
дозы ниже 0,035 Гр сут$^{-1}$. Единственная точка, полученн-
для собак при мощности дозы ниже 10^{-2} Гр·сут$^{-1}$ [189]
жит очень близко к кривой для мыши, имеющей наклон 1,
очевидно, говорит о примерно одинаковой чувствительност
этих двух видов, при таких уровнях мощности дозы, когда
повреждение становится независимым от нее.

Согласно вычислениям D..Grahn, G..A..Sacher, R..A..Lea et al [270] доза облучения 2,5 Зв (т.е. 0,05 Зв, в год, что в настоящее время является предельной эквивалентной доз для профессионального облучения в течение 50 лет работы), данная при хроническом облучении в течение очень продол-тельного периода времени, способна привести к сокращени продолжительности жизни на 15 сут. для средней мыши в расте 100 сут. и примерно на 15 мес. для среднего чел в возрасте 20 лет.

32

Figure 3. From the book by Moskalev Iu.I., Streltsova V.N., 1987; p. 32 [59)

Figure 4 represents another example from the same book [54]. A translation of a fragment from the table, where medical consequences of Chernobyl accident are listed together with corresponding references, is given below in Table 5.

The table contains confusing terminology and doubtful data about morbidity after Chernobyl accident, contradictory to the UNSCEAR report [14], but most remarkable is the second line, where British Islands, Greece and the U.S.A. are presented as territories radiocontaminated after Chernobyl accident.

We have checked the first reference from this line [60]. The statement about 3.3 times incidence increase of childhood leukemia in Wales and Scotland in the years 1987–88 can be related only to the following figures presented in a table (Figure 5) in the letter to the Editor [60]: 1971–80 (15 cases); 1981–86 (9 cases) and the year 1987 (6 cases). Gibson et al. did not put these figures in connection with Chernobyl accident, one year being obviously to short for an induction period of radiation-induced malignancy [61]. Statistically significant increase of leukemia as well as of other malignancies (except for thyroid carcinoma) in consequence of Chernobyl accident was registered neither in West Europe, nor in residents of contaminated areas immediately around Chernobyl [6, 14]. Therefore, this example should be qualified as misinterpretation of statistical data in order to create exaggerated impression about Chernobyl consequences.

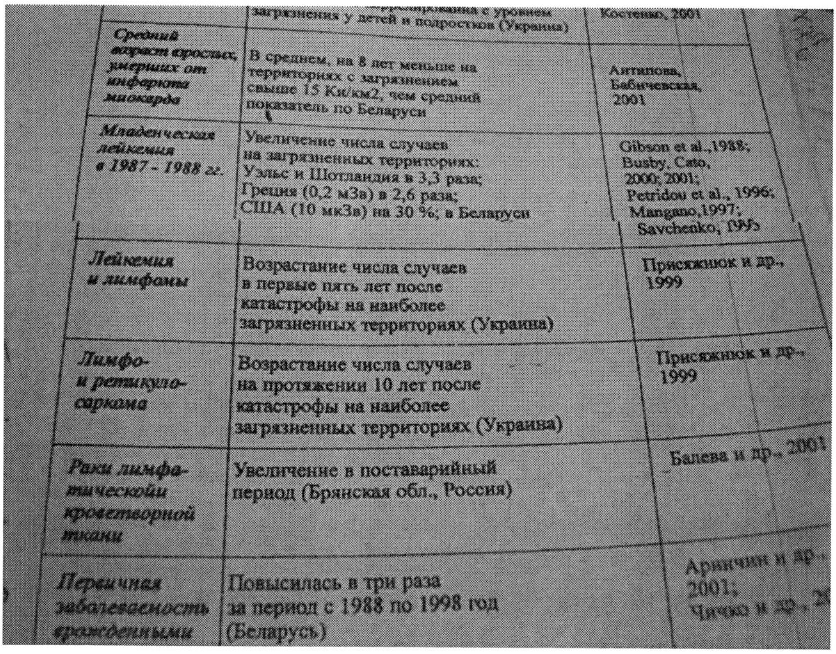

	загрязнения у детей и подростков (Украина)	Костенко, 2001
Средний возраст взрослых, умерших от инфаркта миокарда	В среднем, на 8 лет меньше на территориях с загрязнением свыше 15 Ки/км2, чем средний показатель по Беларуси	Антипова, Бабичевская, 2001
Младенческая лейкемия в 1987 - 1988 гг.	Увеличение числа случаев на загрязненных территориях: Уэльс и Шотландия в 3,3 раза; Греция (0,2 мЗв) в 2,6 раза; США (10 мкЗв) на 30 %; в Беларуси	Gibson et al.,1988; Busby, Cato, 2000; 2001; Petridou et al., 1996; Mangano,1997; Savchenko, 1995
Лейкемия и лимфомы	Возрастание числа случаев в первые пять лет после катастрофы на наиболее загрязненных территориях (Украина)	Присяжнюк и др., 1999
Лимфо- и ретикуло-саркома	Возрастание числа случаев на протяжении 10 лет после катастрофы на наиболее загрязненных территориях (Украина)	Присяжнюк и др., 1999
Раки лимфатической кроветворной ткани	Увеличение в поставарийный период (Брянская обл., Россия)	Балева и др., 2001
Первичная заболеваемость врожденными	Повысилась в три раза за период с 1988 по 1998 год (Беларусь)	Аринчин и др., 2001; Чичко и др., 2[illegible]

Figure 4. From the book by Yablokov, 2002, p. 44 [54)

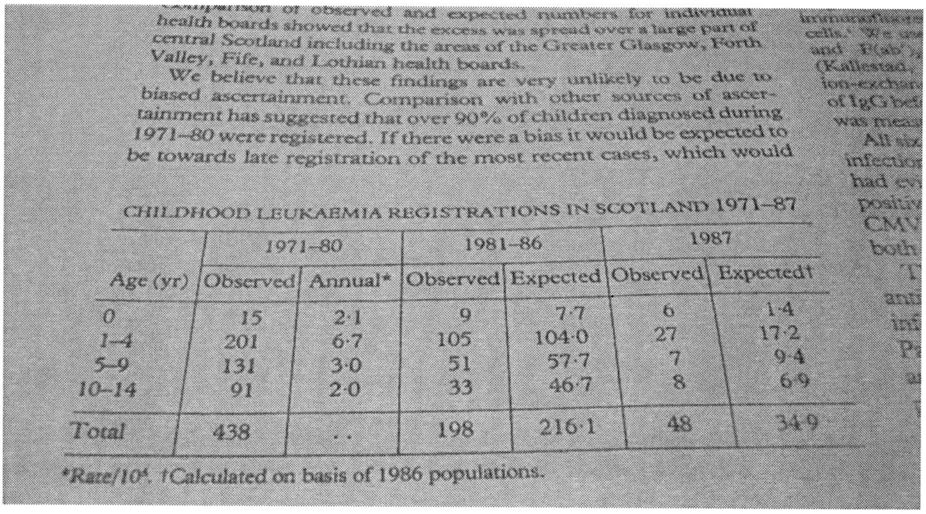

health boards showed that the excess was spread over a large part of central Scotland including the areas of the Greater Glasgow, Forth Valley, Fife, and Lothian health boards.

We believe that these findings are very unlikely to be due to biased ascertainment. Comparison with other sources of ascertainment has suggested that over 90% of children diagnosed during 1971–80 were registered. If there were a bias it would be expected to be towards late registration of the most recent cases, which would

CHILDHOOD LEUKAEMIA REGISTRATIONS IN SCOTLAND 1971–87

	1971–80		1981–86		1987	
Age (yr)	Observed	Annual*	Observed	Expected	Observed	Expected†
0	15	2·1	9	7·7	6	1·4
1–4	201	6·7	105	104·0	27	17·2
5–9	131	3·0	51	57·7	7	9·4
10–14	91	2·0	33	46·7	8	6·9
Total	438	. .	198	216·1	48	34·9

*Rate/10⁵. †Calculated on basis of 1986 populations.

Figure 5. From the Letter to the Editor by Gibson et al., 1988 [60)

Table 5. Medical consequences of Chernobyl accident [54]

Average age of adults died from myocardial infarction	On average, eight years lower in the territories with radiocontamination above 15 Ci/km^2 in comparison with average index for Belarus	Antipova, Babichevskaya, 2001
Leukemia in infants 1987–1988	Increase of the number of cases in contaminated territories: Wales and Scotland 3.3 times; Greece (0.2 mSv) 2.6 times; U.S.A. (10 mkSv) by 30%	Gibson et al., 1988; Busby, Cato 2000, 2001; Petridou et al., 1996; Mangano 1997; Savchenko 1995
Leukemia and lymphoma	Incidence increase in the first five years after the catastrophe in the most contaminated territories (Ukraine)	Prisiazhniuk et al., 1999
Lympho- and reticulosarcoma	Incidence increase in the first ten years after the catastrophe in the most contaminated territories (Ukraine)	Prisiazhniuk et al., 1999
Cancers of lymphatic and hemopoietic tissue	Increased in the post-Chernobyl period (Bryansk region, Russia)	Baleva et al., 2001
Primary incidence of congenital malformations	Increased three times from 1988 to 1998 (Belarus)	Arinchin et al., 2001

Below are some examples of statements from A.V. Yablokov's report [56] at the recent congress of Problems of Radiation Biology and Human Safety, within the framework of the Second St. Petersburg International Ecological Forum. The following items were listed among the causes of proposed underestimation of Chernobyl medical consequences (our comments are in parentheses):

- Irreversible falsification of medical data pertaining to the first years after Chernobyl accident (incidence of radiation induced cancer could not have been elevated in the first years after the accident because of the induction period, which is at least two to three years for leukemia and longer for solid cancers [61])
- Insufficiency of modern scientific knowledge about action specificity even of the main dose-forming radionuclides (action mechanism of radionuclides was studied in detail in numerous experiments, in Russia and in other countries, there are voluminous monographs on this topic)
- Erroneous requirement of statistical significance of correlations (apart from correlation, there are other statistical methods; but statistical data must be significant to allow reliable conclusions)

Using his reputation of a scientist, A.V. Yablokov publishes propagandistic materials in mass media and on the Internet. Some examples of his recommendations concerning nuclear power production in future [55] (with our comments in parentheses):

- "What must be done with energy production, if no nuclear power plants will be built in future? Certainly, to intensify the efforts in the energy saving: it can save 30–40%

of currently used energy... atomic power plants must be replaced by gas-turbine and modern thermoelectric power stations." (Overpriced fuel must be imported for them.)

- "Decentralization of energy production is necessary with reduction of large distance electricity transmission, associated with energy losses up to 30 %." (Decentralized energy production is less efficient and costs in the end more fuel.)
- "All of the above allows concluding that nuclear energy production has no serious perspectives in the foreseeable future. It will be gradually reduced and replaced by other, less dangerous and economically more advantageous energy sources. It is in Russia's interests to accelerate this process." (*Cui prodest scelus is fecit.*)

The publication by A.V. Yablokov [53], best known in the West, is entitled "Radioactive Waste Disposal in Seas Adjacent to the Territory of the Russian Federation." It extracts material from a commission that published a report produced in Russia in 1993. The commission submitted the report to the president of the Russian Federation in February 1993. By presidential decision, this report (after several technical corrections) was open to the public: it is known variously as the "Yablokov Commission Report" or, simply, the "Yablokov Report." During April–May 1993, 500 copies were distributed among governmental agencies inside Russia, and abroad through a net of Russian embassies (from the editor's note in this article).

The text of this publication [53] is overloaded with minor details and represents a typical example of "chattering down" a theme—a method that is used broadly by Russian media to avoid serious discussion. Even the concluding section, "Summary and Conclusions," is vague and unclear. The true conclusion of the "Yablokov Report" is hidden in the middle of the text but can be found and quoted. It is quite clear: "Data from previous research permit us to draw the preliminary conclusion that the LRW (liquid radioactive waste) discharged from facilities of the Northern and Pacific Fleets and the Murmansk Maritime Shipping Line presents no significant radiation danger, either to the population as a whole or to critical population groups (fishermen, residents of coastal areas)." In regard to the solid radioactive waste (SRW), a vague and unconstructive formulation is used: "And until each and every dumped SRW with high activity levels is inspected, no final conclusions concerning them can be drawn." There are reasons to assume that more relevant information about "each and every dumped SRW" can be found in the archives of some institutions in Moscow rather than on the ocean floor.

CONCLUSION

It is not surprising that many scientific and pseudo-scientific publications, overestimating the medical consequences of the Chernobyl accident and elevated background radiation in general, come from the former Soviet Union. The motives are quite clear. In the first phase, heated interest in the Chernobyl theme guaranteed foreign help and international scientific cooperation. Deeper motives are becoming obvious: puffing up Chernobyl hysteria has hindered the advancement of nuclear power production in many countries, thus contributing to higher energy prices. The following tools are used for this purpose: interpretation of spontaneous diseases as radiation-induced; indication of radioactivity or dose levels without confrontation with the natural background radiation; and conclusions about risk or morbidity increase without statistically correct comparison with other regions. In some studies, high figures were obviously caused by non-random material selection or inaccurate morphological assessment. Some studies are based on small samples or singular observations and, although providing no significant information, create an exaggerated impression about the consequences of the accident. Overt misquoting, misinterpretation of statistical data and propagandistic tricks can be found in some publications pretending to be scientific.

References

[1] Grobova, OM; Chernikov, VP. The presence of cesium-137 in the tissue of a lung tumor in someone who cleaned up the aftermath of the accident at the Chernobyl Atomic Electric Power Station (in Russian with English summary). *Ter Arkh.*, 1996, 68(3), 26-30.

[2] Kogan, EA; Cherniaev, AL; Chuchalin; AG; Samsonova, MV; Demura, SA; Sekamova, SM; Zholt, S; Sende, B; Suanova, LA; Ali-Riza, AE. Morphologic and molecular-genetic characterization of lung cancer developing in people who have worked at nuclear facilities and who have lived in Russian territories polluted after the accident at the Chernobyl power plant (in Russian with English summary). *Arkh Patol.*, 1999, 61(1), 22-6.

[3] Derizhanova, IS. The pulmonary-renal syndrome among Chernobyl nuclear accident liquidators. Abstracts of 23rd Congress of the International Academy of Pathology (Nagoya 2000). *Pathology International,* 2000, 50 (Suppl), Abstract P-19-34.

[4] Degtiarova, LV. A possibility of gastric cancer and MALT-lymphomas appearance in persons affected because of Chernobyl accident. Abstracts of 23rd Congress of the International Academy of Pathology (Nagoya 2000*). Pathology International,* 2000, 50 (Suppl), Abstract P-19-31.

[5] Okeanov, AE; Sosnovskaya, EY; Priatkina, OP. National cancer registry to assess trends after the Chernobyl accident. *Swiss Med Wkly.*, 2004, 134(43-44), 645-9.

[6] Chernobyl's Legacy: Health, Environmental and Socio-Economic Impacts and Recommendations to the Governments of Belarus, the Russian Federation and Ukraine. Vienna: IAEA, 2006.

[7] Jargin, SV. Over-estimation of radiation-induced malignancy after the Chernobyl accident. *Virchows Arch.,* 2007 Jul., 451(1), 105-6; author reply 107-8 .

[8] Jargin, SV. Re: Involvement of ubiquitination and sumoylation in bladder lesions induced by persistent long-term low dose ionizing radiation in humans and Re: DNA damage repair in bladder urothelium after the Chernobyl accident in Ukraine. *J Urol.*, 2007, 177(2), 794; author reply 794-5.

[9] Romanenko, A, Morell-Quadreny, L, Ramos, D, Nepomnyaschiy, V, Vozianov, A, Llombart-Bosch, A. Extracellular matrix alterations in conventional renal cell carcinomas by tissue microarray profiling influenced by the persistent, long-term, low-dose ionizing radiation exposure in humans. *Virchows Arch.*, 2006, 448(5), 584-90.

[10] Romanenko, A; Morimura, K; Wei, M; Zaparin, W; Vozianov, A; Fukushima, S. DNA damage repair in bladder urothelium after the Chernobyl accident in Ukraine. *J Urol.*, 2002, 168(3), 973-7

[11] Romanenko, AM; Kinoshita, A; Wanibuchi, H; Wei, M; Zaparin, WK; Vinnichenko, WI; Vozianov, AF; Fukushima, S. Involvement of ubiquitination and sumoylation in bladder lesions induced by persistent long-term low dose ionizing radiation in humans. *J Urol.*, 2006, 175(2), 739-43.

[12] Borovikova, NM; Burlak, GF; Berezhnaya, TI; Varbanets, AN; Tkachenko, NV; Chuprina, SV. Composition of irradiation dose of the population of Kiev after the accident at the Chernobyl atomic power-station. pp. 33-34 In: Results of assessment of medical consequences of the accident at the Chernobyl atomic power-station. *Proceedings of the Scientific and Practical Conference*. Kiev, 1991 (in Russian)

[13] Mould, RF. *The Chernobyl Record. The Definitive History of Chernobyl Catastrophe*. Philadelphia: Institute of Physics, 2000.

[14] UNITED NATIONS. Sources and effects of ionizing radiation. UNSCEAR 2000 Report to the General Assembly, Vol. 1. *Sources and effects of ionizing radiation*. New York: United Nations, 2000. p. 15

[15] Guidelines for drinking-water quality. 3rd Edition. Vol. 1. Recommendations. Radiation aspects, p. 197-209. World Health Organization, 2006. Electronic version for the Web. Available at: http://www.who.int/water_sanitation_health/dwq/gdwq 0506_9.pdf (accessed 23 November 2008)

[16] IARC monographs on the evaluation of carcinogenic risks to humans. *Ionizing radiation*, Part 2. Some internally deposited radionuclides. Lyon: IARC Press 2001,v. 78, pp. 342-343, 232-239.

[17] Romanenko, AM; Morimura, K; Kinoshita, A; Wanibuchi, H; Takahashi, S; Zaparin, WK; Vinnichenko, WI, Vozianov, AF; Fukushima, S. Upregulation of fibroblast growth factor receptor 3 and epidermal growth factor receptors, in association with Raf-1, in urothelial dysplasia and carcinoma in situ after the Chernobyl accident. *Cancer Sci.*, 2006, 97(11), 1168-74.

[18] Isotope Production. In: *Encyclopedic Dictionary of Physics*. Oxford: Pergamon Press, 1961, v. 4, p. 111.

[19] Shirley, VD; Baglin, CM; Frank Chu, SY; Zipkin, J. (Editors). *Tables of Isotopes*. 8th edition. New York: Wiley & Sons Inc., 1996, v. 2, pp. A-151-272.

[20] O'Reilly, PH; Shields, RA; Testa, HJ. *Nuclear medicine in urology and nephrology*. London: *Butterworth & Co.*, 1979, pp. 127-137.

[21] Reuter, VE. The Urothelial Tract: Renal Pelvis, Ureter, Urinary Bladder, and Urethra. In: *Sternberg's Diagnostic Surgical Pathology*. 4th Edition. Mills SE, Carter D, Greenson JK, Overman HA, Reuter VE, Stoler MH (Editors). Philadelphia: Lippincott *Williams & Wilkins,* 2004, v. 2, pp. 2035-2082.

[22] Cardis, E. Current status and epidemiological research needs for achieving a better understanding of the consequences of the Chernobyl accident. *Health Phys.*, 2007,93, 542-546.

[23] Williams, ED. Chernobyl and thyroid cancer. *J Surg Oncol.*, 2006, 94, 670-677.

[24] Holm, LE. Thyroid cancer after exposure to radioactive ^{131}I. *Acta Oncol.* 2006, 45, 1037-1040

[25] Omutov, M; Jargin, SV. The practice of pathology in Russia. Abstracts of the 21st European Congress of Pathology. *Virchows Archiv.*, 2007, 451, 277.

[26] Tronko, MD; Bogdanova, TI; Komissarenko, IV; Epstein, OV; Oliynyk V; Kovalenko, A., et al. Thyroid carcinoma in children and adolescents in Ukraine after the Chernobyl nuclear accident: statistical data and clinicomorphologic characteristics. *Cancer.*, 1999, 86, 149-56.

[27] Demidchik, EP, Tsyb, AF, Lushnikov, EF. *Thyroid carcinoma in children. Consequences of Chernobyl accident* (in Russian). Moscow: *Meditsina,* 1996.

[28] Abrosimov, AIU; Lushnikov, EF; Frank, GA. Radiogenic (Chernobyl) thyroid cancer (in Russian with English summary) *Arkh Patol.*, 2001, 63(4), 3-9.

[29] Chow, LS; Gharib, H; Goellner, JR; Van Heerden, JA. Nondiagnostic thyroid fine-needle aspiration cytology: management dilemmas. *Thyroid.*, 2001, 11, 1147-51.

[30] Abrosimov, AIu. Thyroid carcinoma in children and adolescents of Russian Federation after the Chernobyl accident. Dissertation (in Russian). *Obninsk*, 2004.

[31] Bomash, NIU. *Morphological diagnostics of thyroid diseases* (in Russian). Moscow: Meditsina, 1981.

[32] Kraievski, NA; Smolyannikov, AV; Sarkisov, DS (Editors). Patho-morphological diagnostics of human tumors. *Handbook for physicians* (in Russian). *Moscow, Meditsina*, 1993.

[33] Paltsev, MA; Anichkov, NM. Atlas of human tumour pathology (in Russian). Moscow: *Meditsina*, 2005.

[34] Rosai, J. *Rosai and Ackerman's Surgical Pathology*. Edinburgh: Mosby, 2004, v. 1, pp. 515-594.

[35] Nikiforov, Y; Gnepp, DR. Pediatric thyroid cancer after the Chernobyl disaster. Pathomorphologic study of 84 cases (1991–1992) from the Republic of Belarus. *Cancer*. 1994, 74, 748-66.

[36] Bogdanova, TI; Kozyritskiĭ, VG; Tronko, ND. *Thyroid pathology in children. Atlas* (in Russian). Kiev: Chernobylinterinform, 2000.

[37] Fonseca, E; Soares, P; Cardoso-Oliveira, M; Sobrinho-Simões, M. Diagnostic criteria in well-differentiated thyroid carcinomas. *Endocr Pathol.*, 2006, 17(2), 109-17.

[38] Khurana, KK; Baloch, ZW; LiVolsi, VA. Aspiration cytology of pediatric solitary papillary hyperplastic thyroid nodule. *Arch Pathol Lab Med.*, 2001, 125(12), 1575-8.

[39] Jargin, SV. Examples of plagiarism from the former Soviet Union. Dermatopathology: *Practical & Conceptual*, 2008, 14 (2), 19. Available from: http://derm101.com (Accessed 8 November 2008; search after the author's name)

[40] Russian Pathology: per Scientiam ad Veritatem. Web site: http://www.freewebs.com/ruspat1/; (accessed 23 November 2008)

[41] Parshkov, EM; Sokolov, VA; Proshin, AD; Kurnosova, LV. Characteristics of thyroid cancer incidence in children residing in radio-contaminated areas. In: Chernobyl legacy. Proceedings of the scientific and practical conference. *Kaluga,* 2006. Issue 4, pp. 132-134 (in Russian)

[42] Tsyb, AF; Ivanov, VK; Matvienko, EG. Medical radiological consequences of Chernobyl accident for the population of Russia. In: Chernobyl legacy. Proceedings of the scientific and practical conference. *Kaluga,* 2006. Issue 4, pp. 13-21 (in Russian)

[43] Demidchik, YE; Saenko, VA; Yamashita, S. Childhood thyroid cancer in Belarus, Russia, and Ukraine after Chernobyl and at present. *Arq Bras Endocrinol Metabol.*, 2007, 51(5), 748-62.

[44] Reiners, C; Demidchik, YE; Drozd, VM; Biko, J. Thyroid cancer in infants and adolescents after Chernobyl. *Minerva Endocrinol,* 2008, 33(4), 381-95.

[45] Kogan, EA; Sagindikova, GS; Sekamova, SM; Jack, G. Morphological, cytogenetic and molecular biological characteristics of lung cancer in persons exposed for a long time to radionuclide radiation pollution in the Semipalatinsk region of Kazakhstan (in Russian with English summary) *Arkh Patol.*, 2002, 64(5), 13-8.

[46] Cotran, RS; Kumar, V; Robbins, SL. *Robbins' Pathologic Basis of Disease.* Philadelphia: W.B. Saunders Co., 1994.

[47] Sagindikova, GE; Dissertation. Moscow IM. *Sechenov Medical Academy*, 2001.

[48] Brenner, DJ; Doll, R; Goodhead, DT., et al. Cancer risks attributable to low doses of ionizing radiation: assessing what we really know. *Proc Natl Acad Sci U.S.A.* 2003, 100(24), 13761-6.

[49] Cohen, BL. Test of the linear-no threshold theory: rationale for procedures. *Dose Response,* 2006, 3(3), 369-90.

[50] Prekeges, JL. Radiation hormesis, or, could all that radiation be good for us? *J Nucl Med Technol.*, 2003, 31(1), 11-7.

[51] Ghiassi-nejad, M; Mortazavi, SM; Cameron, JR; Niroomand-rad, A; Karam, PA. Very high background radiation areas of Ramsar, Iran: preliminary biological studies. *Health Phys.*, 2002, 82(1), 87-93.

[52] Balaram, P, Mani, KS. Low dose radiation--a curse or a boon? *Natl Med J India.* 1994, 7(4), 169-72.

[53] Yablokov, AV. Radioactive waste disposal in seas adjacent to the territory of the Russian Federation. *Mar Pollut Bull.*, 2001, 43(1-6), 8-18.

[54] Yablokov, AV. *A mythos about safety of small radiation doses.* Moscow: Center for Russian Environmental Policy, 2002 (in Russian)

[55] Yablokov, AV. Stroitelnye Issledovaniya (Construction Research). Moscow: Center for Russian Environmental Policy, 2000 (in Russian). Available at: http://www.vintagearabians.com/page10.html, accessed on November 23, 2008.

[56] Yablokov, AV. 2008. Consequences of Chernobyl catastrophe for public health – criticism of the IAEA/WHO position. Report at the Congress "Problems of radiation biology and human safety" (July 3, 2008, Saint-Petersburg, Russia (http://ecoforum2008.com/en/preliminary/rad/). An abstract (in Russian) is published in: *Vestnik Rossiiskoi voenno-meditsinskoi akademii.* 23(3), Suppl 2 (Part 1), 241-242. Evered, D; Lazar, P. Misconduct in medical research. *Lancet;* 1995, 345(8958), 1161-2.

[57] Scientific Misconduct: Misquoting. Website "Russian Pathology" Available at: http://www.freewebs.com/ruspat1/apps/photos/, accessed on November 23, 2008.

[58] Moskalev, Iu.I., Streltsova, VN. Radiation biology. Moscow: VINITI, 1987; p. 32 (in Russian)

[59] Gibson, BE; Eden, OB; Barrett, A; Stiller, CA; Draper, GJ. Leukemia in young children in Scotland. *Lancet.*, 1988, 2(8611), 630.

[60] Ron, E. Ionizing radiation and cancer risk: evidence from epidemiology. *Radiat Res.*1998,150(5Suppl),S30-41.

In: Environmental Regulation: Evaluation, Compliance . . . ISBN: 978-1-60741-645-6
Ed: Diederik Meijer and Fillipus De Jong

Chapter 2

APPLYING MOLECULAR BIOLOGY TECHNIQUES FOR EVALUATING OIL FIELD MICROBIAL BIODIVERSITY AND ENVIRONMENT

Shao Hong-Bo[1,2,4*]***, Zhuang Wen***[2] ***and Ni Fu-Tai***[3*]

[1]Binzhou University, Binzhou 256603, China
[2]Institute for Life Sciences, Qingdao University of Science & Technology, Qingdao 266042, China
[3]College of Life Sciences, Jilin Normal University,Siping 136000,China
[4]State Key Laboratory of Soil Erosion and Dryland Farming on the Loess Plateau, Institute of Soil and Water Conservation, Chinese Academy of Sciences and Ministry of Water Resources, Northwest A&F University, Yangling 712100, China

ABSTRACT

With more and more attention given to the study of microorganisms in oil fields in recent years, a variety of molecular biology and biotechnology techniques are also applied. This chapter describes the most commonly-used molecular biology techniques and related biotechnological advances in oil fields research at present, which[1] are based mainly on 16s rRNA, such as 16S rRNA sequencing, Denaturant Gradient Gel Electrophoresis (DGGE), Terminal Restriction Fragment Length Polymorphism (T-RFLP) and so on. The chapter also points out their advantages and disadvantages, and introduces some of the microorganisms of greatest concern in research.

Key words: biotechnology, molecular biology techniques, oil field, 16s rRNA, microbial biodiversity, environment

---

[1] *Corresponding Author: Hong-Bo Shao, Institute for Life Sciences, Qingdao University of Science & Technology, Qingdao 266042, China, Tel: +86-532-84023984, Email: shaohongbo@qust.edu.cn (H.B.Shao) and nifutai@163.com(F.T.Ni)
**Shao Hong-Bo and Zhuang Wen contributed equally to the work in this chapter.

1. Introduction

In oil fields with long-term water development, a relatively stable microbial community is established in formation water. The vast range of substrates and metabolites present in hydrocarbon-impacted soils surely provides an environment for the development of a quite complex microbial community (Jonathan D. Van Hamme et al., 2003). In formation water, under relatively stable conditions, the types and quantities of microbial communities are unchanged. In the deeper strata, due to high temperature and pressure, the microbial community structure is relatively simple, while in shallower formations where the temperature is below 100°C the microbial community is more complicated and present in large quantities, mainly due to hydrocarbon-degrading bacteria, denitrifying bacteria, methane-producing bacteria, sulfate-reducing bacteria, iron bacteria, total growth bacteria and so on.

For long-term survival in the formation environment, these bacteria exhibit strong adaptability. The existence of these bacteria, on one hand, has a positive role in enhancing oil recovery; on the other hand, they may cause the corrosion and plug of injection wells and oil wells. To this end, microbiologists are trying to find methods that can inhibit bacteria not conducive to oil production safety, and at the same time promote the growth and reproduction of bacteria conducive to recovery.

For a long time, research into the beneficial or harmful microorganisms in the oil reservoir has diminished, whether for the role of oil production (Bass et al., 1997) or for metal corrosion. At present, the use of microorganisms to raise oil output in oil reservoirs (Microbial-Enhanced Oil Recovery [MEOR]) is popular around the world and is developing rapidly. At the same time, Biocompetitive Exclusion technology (BCX technology) has been successfully tested in some oil fields. Therefore, in oil production projects, for the use of microbial competition exclusion technology to control the occurrence and development of hydrogen sulfide, or for the use of microbial-enhanced oil recovery technology to increase oil output, the analysis of bacterial communities in reservoirs is very necessary.

2. Types of Important Microbiological Resources

2.1. Sulfate-Reducing Bacteria

Sulfate-reducing bacteria (SRB) are anaerobic microorganisms that have been found to be involved with numerous Microbiologically Influenced Corrosion (MIC) problems affecting a variety of systems and alloys. They can survive in an aerobic environment for a period of time until finding a compatible environment. In offshore oil production, seawater is commonly injected into the reservoirs to enhance secondary oil recovery. The sulfate-rich seawater stimulates growth of sulfate reducing bacteria (SRB) in the reservoirs with subsequent H_2S production. This biogenic H_2S production, designated "reservoir souring", is of major concern to the oil industry, as H_2S is toxic and corrosive, increases sulfur content in oil and gas, and may lead to reservoir plugging (Myhr et al., 2002). The most common strains exist in the temperature range from 25 to 35°C, although there are some that can live at temperatures of 60°C. They can be detected through the presence of black precipitates in the liquid media or surface deposits, as well as by their characteristic hydrogen sulfide smell (Lane, 2005). One report by Iverson shows that in the United States, 77% corrosion of oil wells are caused by SRB, the characteristic of which is pit corrosion; the corrosion rate of steel can increase to 15 fold. Hence, corrosion brought on by SRB is one of the most important problems to be solved (Videla, 1996; Kim et al., 2001; Wang et al., 2004).

2.2. Nitrate-Reducing Bacteria (NRB)

NRB can be classified on the basis of the electron donors that they use (Eckford et al., 2002). They can be chemolithotrophs that use inorganic compounds such as sulfide, thiosulfate or ferrous sulfide as electron donors (Gevertz et al., 2000; Kuenen, 1989; Sublette et al., 1994), or chemoorganotrophs (heterotrophs) that use organic compounds as electron donors (Zumft, 1992). In the presence of organics such as nitrate and volatile fatty acids, NRB multiply rapidly. Through metabolism they can transform the nitrate into nitrite, and further into gases such as N_2 and N_2O. They also can remove sulfide in the system to inhibit the growth of sulfate-reducing bacteria by survival competition, and inhibit the formation of new sulphides and reduce corrosion on production equipment by toxic gases such as H_2S, and damage on the formation by insolubles such as Fe (Sperl; Dennis, 1998).

2.3. Iron/Manganese-Oxidizing Bacteria

Iron- and manganese-oxidizing bacteria have been found in conjunction with MIC, and are typically located in corrosion pits on steels. Some species are known to accumulate iron or manganese compounds resulting from the oxidation process. High concentrations of manganese in biofilms have been attributed to the corrosion of ferrous alloys, including pitting of stainless steels in treated water systems. Iron tubercles have also been observed to form as a result of the oxidation process (Lane, 2005).

2.4. Hydrocarbon-Degrading Bacteria

Hydrocarbon-degrading bacteria (HDB) refer to the bacteria groups that can use petroleum hydrocarbons as a substrate of growth. This type of bacteria are most abundant in water injection wells and near the bottom zone, and are the start of the food chain link of microorganisms in the water injection oil layer. HDB can produce catabolic enzymes through metabolism, crack heavy hydrocarbons and paraffin, reduce oil viscosity, and improve the flow of oil so as to raise oil recovery. In addition, HDB can produce surfactants, polymers, organic acids, alcohol and carbon dioxide by metabolism, which are conducive to oil flooding. The majority of HDBs are aerobic bacteria; injecting air containing water with nutrients into formation can greatly stimulate the growth of these bacteria so as to enhance oil recovery.

2.5. Total Growth Bacteria

Total growth bacteria (TGB) comprise all kinds of heterotrophic bacteria, that is, bacteria living on organic material. Some bacteria can output products that improve the flow of oil by metabolism. Viscous substances and metabolites produced by a certain group of anaerobic bacteria as well as organism accumulation can have an impact on the formation. TGB are often referred to as aerobic-heterotrophic bacteria; some of these bacteria in the process of growth and reproduction can produce a large number of viscous substances. The substances attach to the pipelines and equipment, form scales, plug the water injection wells and filters and, once entering the formation, will cause plugs; at the same time they will create oxygen concentration cells and cause corrosion, sometimes forming local anaerobic environments which are suitable for the growth of SRB, so as to increase corrosion. Therefore, this type of bacteria is an important control indicator in oil-injected water.

3. Molecular Biology Methods

In order to achieve Bio-Competitive Exclusion technology, we must understand the composition of microbial sources in oilfield. Traditional culture techniques have yielded valuable information about microbial interactions with hydrocarbons in the environment. However, one must keep in mind that only a small fraction of microorganisms can currently be cultured from environmental samples, and even if a microorganism is cultured, its role in a community and contribution to ecosystem function are not necessarily revealed (Van Hamme et al., 2003). There are a large number of uncultured microbes and cultured independent microbes in oil reservoirs which can not grow on pure culture of nutritious medium in the laboratory. At the same time ,microorganisms inevitably will be placed in a new artificial manipulation environment deviating from their original small environment, which will change the original structure of microbial communities, hence, micro-organisms may be deviated from the original nature of the physical habits of the population, even depart from their original genotype combination. Therefore, any use of traditional microbial technology to a correct understanding of the microbial ecosystems will face considerable obstacles, a lot of microbial resources of great practical value (uncultured microbes contain about 99 percent) are also buried, leading to research level of microbial diversity and ecological has lagging behind other communities in oil reservoir.

Culture-independent approaches to microbial community analyses have recently enjoyed a surge in popularity as new techniques have been developed and are available in most major research institutions. Molecular descriptions of microbial communities now dominate the literature in all areas of microbial ecology, not just petroleum microbiology (Van Hamme et al., 2003). Since 1985, according to Pace et al., using DNA sequencing technology to study microbial ecology and evolutionary issues, the research of microbial diversity has entered a new stage, and methods and techniques of the molecular ecological technology of microorganisms have gradually developed. Internationally, a large number of research and development practices have proven that using molecular ecological technology to study microorganisms cannot be restricted by whether the microorganism can be isolated and cultured, and whether microorganisms are lively in the laboratory, so the complex microbial community structure can be analyzed quickly, accurately, and thoroughly.

At present, microbial molecular ecology research is focused on 16S rRNA sequencing and analysis, DGGE, T-RFLP, FISH, and PCR and so on. Here are several major research methods in molecular biology and their applications in oil microbial biodiversity research.

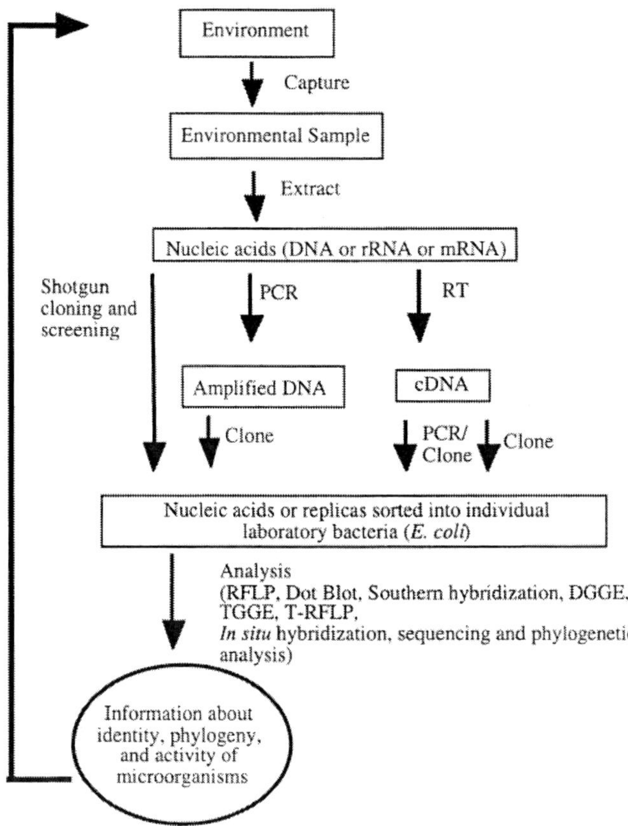

Figure 1. Steps toward nucleic-acid analysis of naturally occurring microorganisms. cDNA, complementary DNA; DGGE, denaturing gradient gel electrophoresis; TGGE, temperature gradient gel electrophoresis; PCR, polymerase chain reaction; RFLP, restriction fragment length polymorphism; RT, reverse transcriptase; T-RFLP, terminal restriction fragment length polymorphism (Madsen EL 2000).

3.1. 16S rRNA Sequencing and Analysis

16S rRNA sequencing and analysis is an important way to infer phylogeny of bacteria and their evolutionary relationship, using comparative analysis of the homology of 16S rRNA sequences of different bacteria. The basic principle is as follows: extract the 16S rRNA gene fragments from microbial samples; next, cloning, sequencing or digestion; the probe hybridizations are used to gain 16S rRNA sequence information and then compare the 16S rRNA sequence data in the database or other data; calculate evolutionary distance between them by the differences in sequence; and determine their position in the evolutionary tree to determine the types of microbes that may exist in the samples.

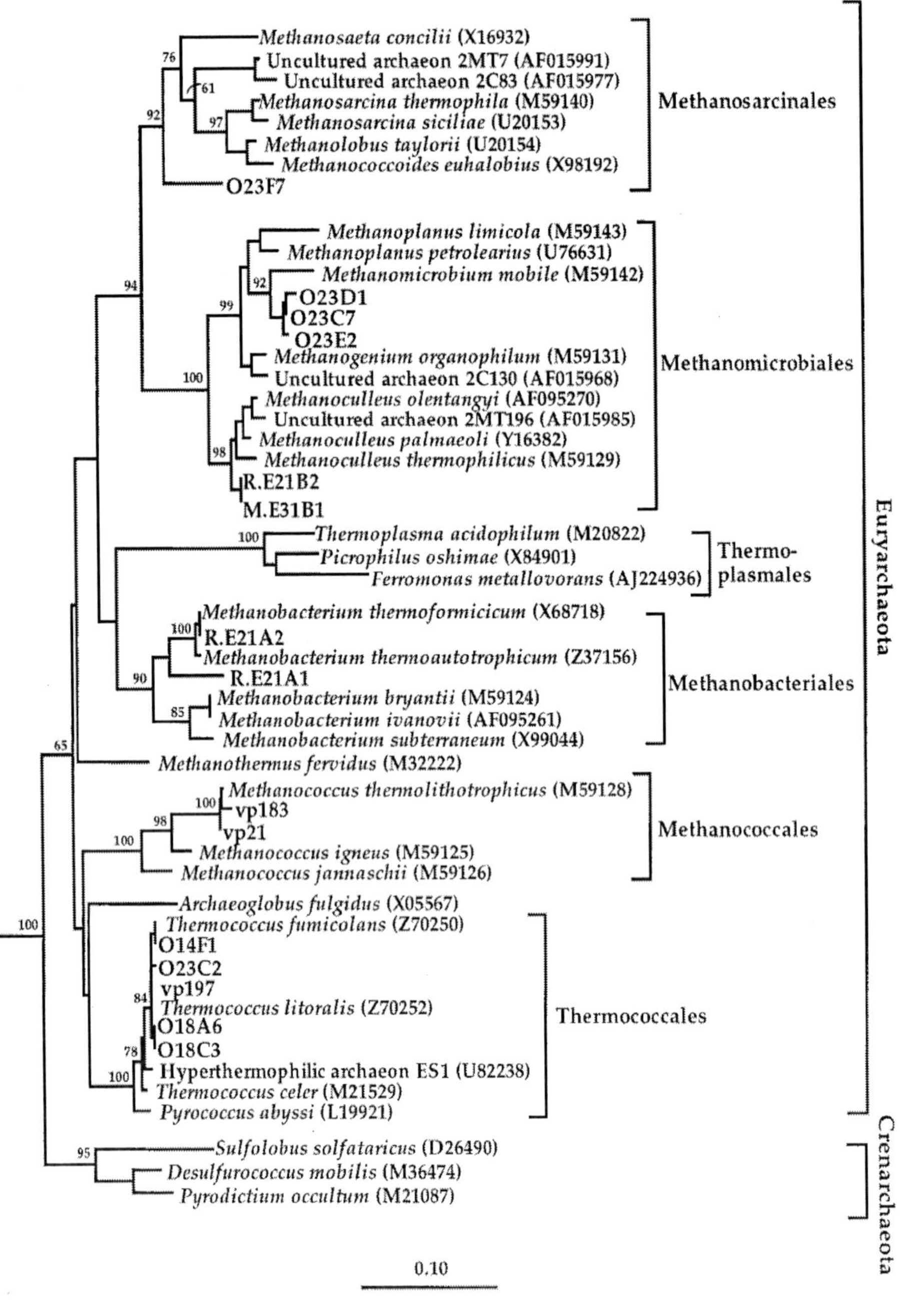

Figure 2. Example of phylogenetic tree. Phylogenetic tree of the archaeal domain and archaeal 16S rDNA phylotypes (Orphan et al., 2000). Bootstrap values (n=1,000 replicates) of $\geq$50 are reported as percentages. The scale bar represents the number of changes per nucleotide position.

Table 1. Example of Closest relatives of archaeal phylotypes from 16S rDNA libraries (Orphan et al., 2000)

Division (% representation)	Type sequence	No. of RFLP types	No. of clones	Closest cultivated species	Source	% Sequence similarity
Methanomicrobiales (96%)	O2 3C7[a]	4	145	*Methanoplanus*	Oil	95.7–
				petrolearius		96.1
				Methanomicrobium	Cow	
Methanosarcinales (6%)				*mobile*	rumen	93.9–
	O2 3F7[a]	1	7	*Methanosarcina*	Sewage	94.4
Thermococcales (1.3%)				*thermophila*		
	O2 3C2[a]	1	2	*Thermococcus*	Oil	87.6
Thermococcales (8.8%)				*litoralis*		
	O1 8A6[b]	5	14	*Thermococcus*	Oil	99.1
				litoralis		
						99.4–
						99.9

[a] 16S rRNA *E. coli* numbering (20 to 958).
[b] 16S rRNA *E. coli* numbering (519 to 1,390).

For example, database searches with 16S rDNA sequences were conducted using the BLAST program and the GenBank database. The profile alignment technique of ClustalW version 1.7 was used to align the sequences, and the alignments were refined by visual inspection; secondary structures were considered for the refinement analysis. A phylogenetic tree was constructed by the neighbor-joining method using the njplot software in ClustalW, version 1.7. Nucleotide positions at which any sequence had an ambiguous base were not included in the phylogenetic calculations. Checks for chimeric sequences were conducted by using the chimera check in the RDP database (Watanabe et al. 2002).

One report by Orphan et al. (2000)shows that two 16S rRNA gene libraries were generated from total community DNA collected from high-temperature, sulfur-rich oil reservoirs in California, using either archaeal or universal oligonucleotide primer sets. Sequence analysis of the universal library indicated that a large percentage of clones were highly similar to known bacterial and archaeal isolates recovered from similar habitats.

Gevertz et al. (2000) isolated two novel nitrate-reducing, sulfide-oxidizing Bacteria strains CVO and FWKO B, from produced brine at the Coleville oil field in Saskatchewan, Canada. The present 16s rRNA analysis suggests that both strains are members of the epsilon subdivision of the division *Proteobacteria*, with CVO most

closely related to *Thiomicrospira denitrifcans* and FWKO B most closely related to members of the genus *Arcobacter*.

Li Hui et al. (2007) analyzed the diversity of an archaeal community in the water from a continental high-temperature, long-term water-flooded petroleum reservoir in Huabei Oilfield in China. The archaea were characterized by their 16S rRNA genes. Phylogenetic analysis of these sequences indicated that the dominant members of the archaeal phylotypes were affiliated with the Methanomicrobiales; some had been previously isolated from a number of high-temperature petroleum reservoirs worldwide, so they are speculated to exhibit adaptations to the environment and be the common habitants of geothermally-heated subsurface environments.

3.2. Denaturant Gradient Gel Electrophoresis (DGGE)

DGGE is a method to separate DNA samples multiplied by PCR. DNA samples are pulled through a gel, containing a gradient of denaturant using an electrical field. Depending on which base pairs the DNA contains, it stops at specific denaturant concentrations. A sample containing a number of bacterial strains thus gives a corresponding number of bands on a DGGE gel. The position and number of bands may itself be sufficient information in comparing different samples. The bands may also be cut out and the base pair sequence decoded by direct gene sequencing. Denaturing gradient gel electrophoresis (DGGE) has been used to resolve PCR-amplified regions of genes coding for 16S rRNA (16S rDNA) based solely on differences in nucleotide sequence (Øvreas et al. 1997; Ferris et al., 1996). This has proven to be a simple approach to obtain profiles of microbial communities that can be used to identify temporal or spatial differences in community structure or to monitor shifts in structure that occur in response to environmental perturbations (Øvreas et al. 1997; Ferris et al., 1996; Muyzer et al., 1993; Wawer et al., 1995). Moreover, since each DNA fragment in the profile is likely to be derived from one (or few) phylogenetically distinct populations, one can readily obtain an estimate of species number and abundance based on the number and intensity of amplified fragments in the profile. It has also been possible to infer the phylogeny of community members by DNA sequence analysis of amplified fragments after they have been excised (Øvreas et al., 1997; Muyzer et al., 1993; Muyzer et al., 1995; Rolleke et al., 1996; Teske et al., 1996).

Wang Jun et al. (2008) monitored changes in exogenous bacteria and investigated the diversity of indigenous bacteria during a field trial of microbial enhanced oil recovery using DGGE. DGGE profiles indicated that the exogenous strains were retrieved in the production water samples and indigenous strains could also be detected. Sequence analysis of the DGGE bands revealed that Proteobacteria were a major component of the predominant bacteria. Their experiment confirms that DGGE analysis was a successful approach to investigate the changes in microorganisms used for enhancing oil recovery.

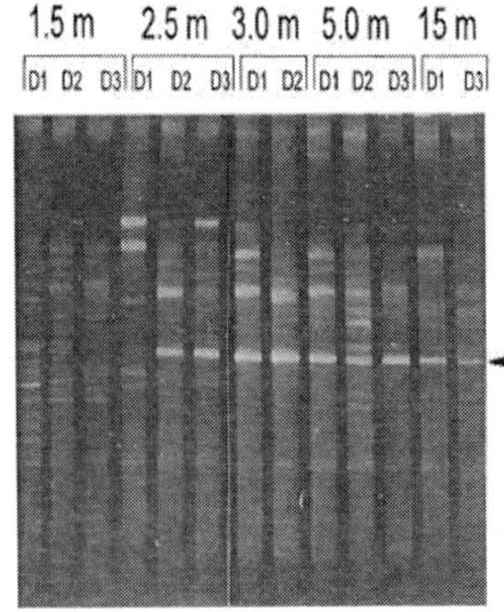

Figure 3. Example of a DGGE gel on DNA extracted from Lake Sælenvannet (Øvreas et al., 1997). DGGE analysis showing variations in time and space of bacterioplankton community structure. Bacterial primers were used for PCR amplification. Five different depths and three different sampling times were processed. D1, 22 March 1995; D2, 29 January 1996; D3, 5 March 1996. The arrow labeled 17 indicates the fragment with high sequence similarity to *C. phaeovibrioides* DNA.

Forney et al. applied the DGGE method to determine the relative genetic complexity of microbial communities at different depths in the meromictic Lake Sælenvannet. The technique was also used to monitor community changes in space and time (Øvreas et al. 1997). The results show that among the dominant populations were representatives related to *Chlorobium phaeovibrioides*, chloroplasts from eukaryotic algae, and unidentified *Archaea.*

3.3. Fluorescent In Situ Hybridization (FISH)

The genetic material (DNA, RNA) inside bacteria is unique for all individual species. Using, e.g., the sequences of base pairs in the 16S rRNA, probes can be designed to target individual groups, generas or strains of bacteria. Adding a fluorophore (fluorescent due) to the probe enables detection of cells to which the specific gene probe has attached. Though the preparation steps are different, the cells are counted as in direct bacterial counts. Using different fluorophores it is possible to stain more than one group at a time, and including a general stain like DAPI it is possible to obtain quantitative information (% specific group of total population) (Højris et al.).

Kleikemper et al. (2002) characterized the SRB population in a PHC-contaminated aquifer by using FISH. For in situ hybridization, they used the indocarbocyanine (Cy3)-labeled 16S rRNA oligonucleotide probes (all purchased from MWG Biotech, Ebersberg, Germany) EUB338 to target Bacteria, Arch915 for Archaea, SRB385 plus SRB385-Db for SRB, DSV698 plus DSV1292 for *Desulfovibrio*, DSB985 for *Desulfobacter*, and probe 660 for *Desulfobulbus*. The result showed that a large fraction of suspended bacteria hybridized with SRB-targeting probes SRB385 plus SRB385-Db (11 to 24% of total cells).

3.4. Terminal Restriction Fragment Length Polymorphism (T-RFLP)

T-RFLP is also known as the 16 S rRNA gene terminal restriction fragment (terminal restriction fragment, TRF) analysis technology, which is a recently emerging molecular biology technique used in microbial polymorphism research. Compared to other methods of molecular biology, there are some distinct advantages in T-RFLP technology. Sequence database with direct reference value, that is all size of end fragments obtained from the digestion can be compared to end fragments in sequence database, hence, phylogeny can be inferred. Compared with DGGE relying on electrophoresis system, the results of DNA sequencing technology is more reliable. Analysis of capillary gel electrophoresis in T-RFLP is more quickly, and the output of the results are in the form of dates. Distinct advantages of T-RFLP make it an ideal community comparative analysis method. Therefore, T-RFLP has been more drawing attention of researchers.

Yuan et al. (2007) used T-RFLP technique to analyze the microbial diversity of an injection well (S122ZHU) and three related production wells (S1224, S1225 and S12219) in the Sheng Li oilfield in China. The Shannon-Wiener Diversity index, based on the T-RFLP profiles, indicated that the bacterial and archaeal species richness in the injection well was higher than those of the production ones. This study indicates that T-RFLP is useful for analysis of the microbial diversity in petroleum reservoirs.

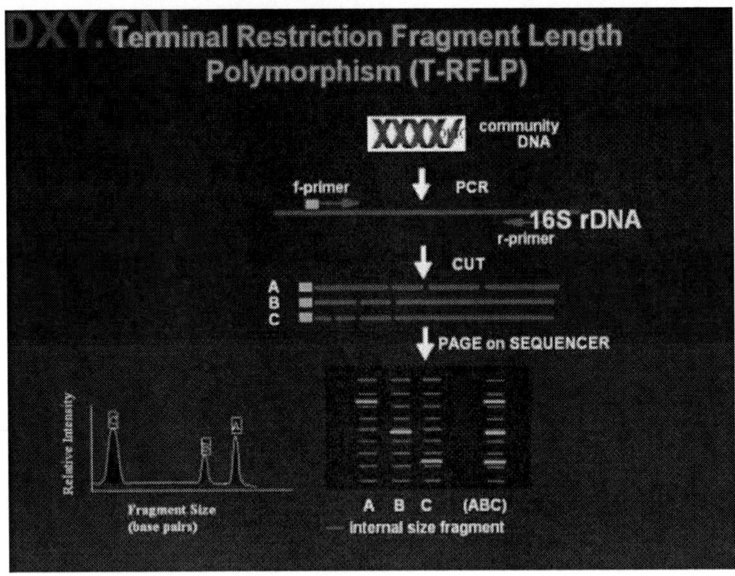

Figure 4. The flow chart of T-RFLP analysis technology (image from http://www.dxy.cn)

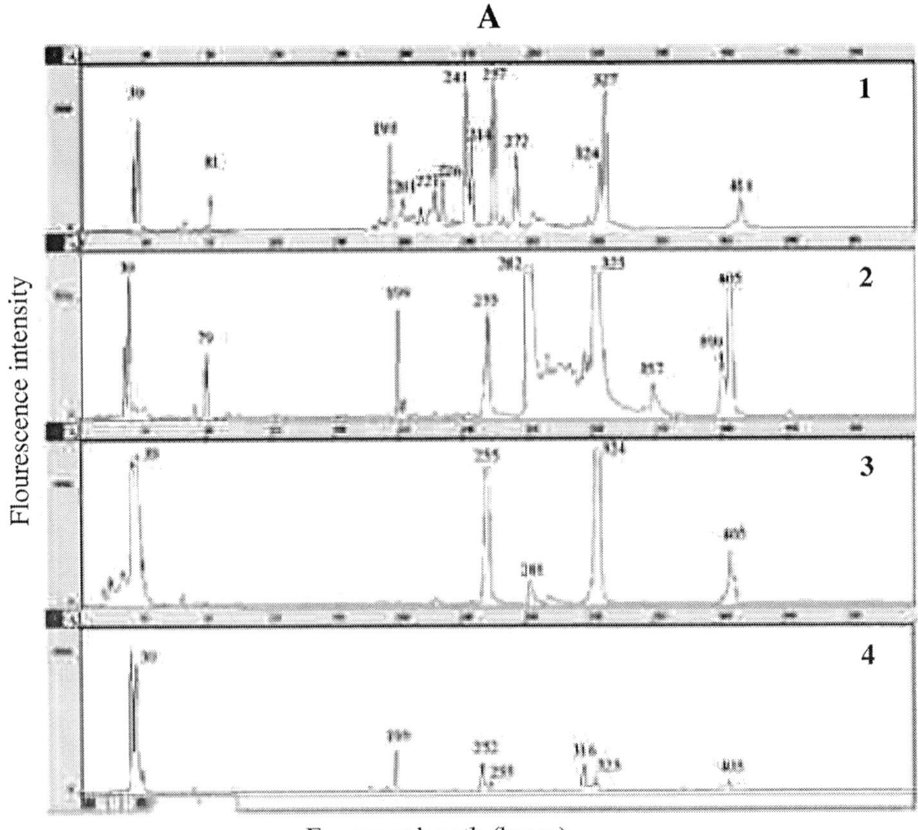

Figure 5. Example of T-RFLP profiles in the research by Yuan et al. (2007). *Hae* β-digesting T-RFLP profiles for bacteria.

3.5. Quantitative Dot Blot

Information on the abundance of a specific DNA sequence can be obtained by using quantitative dot blot for the extraction of DNA from environmental samples. The main principle is as follows: with a separate hybridization of a specific probe (such as specific 16S rRNA gene probe) or a universal probe (such as the total 16S rRNA gene probe) to the total DNA separated from environmental samples, the relative abundance of the specific DNA (such as the 16 S rRNA gene sequences) can be determined by the ratio of two hybridization signals of the two types of probes, so as to reflect the number of cells of a specific microbe or the relative physical activity of specific populations.

3.6. Quantitative Polymerase Chain Reaction (qPCR)

Genetic material may be extracted from the total population in a sample. Though such extracts contain information on all cells present, the concentrations of genetic material is too low to be analysed. The polymerase chain reaction enables exponential multiplication of the DNA, increasing the concentration to measurable amounts (~10^9 gene copies). With a few adjustments the PCR method can be made quantitative, called Real Time Quantitative PCR (in short, qPCR). In the qPCR method a fluorescent signal is measured after each cycle in the PCR. This fluorescent signal corresponds with the amount of DNA produced in the PCR, which again corresponds with the initial number of specific bacterial cells in the sample. Most qPCR assays have a linear range spanning more than six orders of magnitude regarding DNA concentration. The method can detect very few (theoretically only one) gene copies in the initial sample. However, the detection limit (sensitivity) and target specificity has to be carefully evaluated in each new qPCR application (Højris; Skovhus et al., 2004).

3.7. Micro Auto Radiography (MAR)

To determine if a bacterium is capable of utilizing a specific substrate (e.g., glucose or acetate), the bacteria can be exposed to a medium containing a radioactive variation of the substrate in question. If the bacteria can utilize the substrate, it becomes slightly radioactive. Within a complex sample of various strains some bacteria may become radioactive and some not. The sample can be developed like a photographic film using silver grains as indicators for radioactivity, and an image of cells capable and not capable of utilizing the specific substrate may be obtained (Højris; Nielsen et al., 2002; Nielsen et al., 2002).

3.8. Bio-Chip (Biochip)

Thanks to the integration of DNA blotting hybridization technology, a revolution is taking place in molecular biology techniques, i.e., biochip technology, also known as gene chip or DNA arrays. Biochips are composed of tens of thousands of gene probes closely aligned as networks. Through combining the DNA fragments whose base sequences have been known to mark DNA or RNA that possess complementary sequences of base pairs, we can determine the corresponding categories, and infer microorganism groups in environmental samples. The current view of the 16S rRNA gene chip is still in the stage of basic research; however, with the wider use of 16S rRNA/DNA technology, such products in the market is an inevitable trend.

4. Issues and Future Prospects

The nature and diversity of bacteria in oil field ecosystems is still poorly understood. However, recent incorporation of molecular methods has allowed a broader characterization of microbial assemblages in this type of environment (Sette et al., 2007). However, any method used to study the composition of a microbial community has its limitations. Madsen (2000) reviewed many of the major nucleic acid-based methods used for characterizing naturally occurring microorganisms, and he listed limitations for each procedure (Table 2).

During 16S rRNA sequencing and analysis, DNA of microorganisms may have been missed due to PCR biases when using community DNA, such as preferential amplification, or differential lysis efficiency may have occurred when purifying DNA from the samples prior to 16S rRNA gene library assembly.

A study by Watanabe et al. (2002) shows that methods employed in the molecular ecological approaches largely affect the results obtained. In one report Kazuya Watanabe pointed that they did not use FISH, because a possible bias that causes the underestimation of slow-growing bacteria has been suggested. Besides, in the present study, FISH with the *Bacteria*- and *Archaea*-specific probes detected only 60% in total of the DAPI-stained cells; in addition, a large portion of labeled cells exhibited weak signals.

Lopez et al. (2003) point out that, while using PCR-DGGE to examine the bacteria in wine fermentations, they noted that several commonly-used PCR primers for amplifying bacterial 16S rDNA, along with coamplified yeast, fungal, or plant DNA were present in samples. Amplification of nonbacterial DNA can result in a masking of bacterial populations in DGGE profiles. Coamplification of nonbacterial DNA is problematic since it can result in an overestimation of the bacterial content of any particular niche. Moreover, competition between bacterial and nontarget templates during PCR may mask lower bacterial populations. This work demonstrates the importance of testing purportedly bacterium-specific PCR primers on potential eukaryotic DNAs that might co-purify with bacterial DNA in environmental samples prior to embarking on a detailed analysis.

Wang et al. (2008) also pointed out the shortcomings of the use of PCR-DGGE. First, PCR-DGGE profiles could only detect the predominant population in the environmental samples. Second, the reported sensitivity of DGGE was 1% of the template DNA. Third, many problems might arise during sample collection, DNA extraction, PCR amplification or DGGE steps (gel resolution and gel staining). Only microorganism populations with high concentrations could be detected by DGGE fingerprinting.

Although the PCR and DNA extraction used in investigating the microbial ecosystems would introduce biases (Polz et al., 1998; Suzuki et al., 1996; Von Wintzingerode et al., 1997), they are the primary sources of information available to assess the phylogenetic richness and complexity of microbial communities. Additionally, compared to other molecular techniques, the RNA approach has a great advantage: the generation of sequence data can be used to design group-specific probes and primers for further studies such as microarray and real-time PCR, which have not been used to characterize microbial diversity in the petroleum reservoir (Li et al., 2007).

The use of T-RFLP and DGGE with microbial community relations in reservoirs arose from the traditional research methods at the molecular level. Research shows that DGGE technology can be used to find advantageous bacteria groups and changes in bacterial diversity, and through T-RFLP technology we can find the main species in the reservoir and a situation of abundance. The combination of the two kinds of methods can truly reflect the relationship between a microbial community and a reservoir.

Some scholars suggest that a more comprehensive assessment of microbial diversity in oil, and probably other environments, can be obtained by using a combination of culture- and molecular-based techniques rather than by using either method alone (Sette et al., 2007), based on both molecular and cultivation techniques, allowing researchers insights into the microorganisms that might be involved in the biogeochemical transformations that take place in these environments.

Table 2. A summary of current techniques used to study complex microbial ecosystems (Zoetendal et al., 2004)

Methods	Uses	Limitations
Cultivation	Isolation; "the ideal"	Not representative; slow & laborious
16S rDNA sequencing	Phylogenetic Identification	Laborious; subject to PCR biases
DGGE/TGGE/TTGE	Monitoring of community/ population shifts; rapid comparative analysis	Subject to PCR biases; semi-quantitative; identification requires clone library
T-RFLP	Monitoring of community shifts; rapid comparative analysis; very sensitive; potential for high throughput	Subject to PCR biases; semi-quantitative; identification requires clone library
SSCP	Monitoring of community/population shifts; rapid comparative analysis	Subject to PCR biases; semi-quantitative; identification requires clone library
FISH	Detection; enumeration; comparative analysis possible with automation	Requires sequence information; laborious at species level
Dot-blot hybridization	Detection; estimates relative abundance	Requires sequence information; laborious at species level
Quantitative PCR	Detection; estimates relative abundance	Laborious
Diversity microarrays	Detection; estimates relative abundance	In early stages of development; expensive
Non-16S rRNA profiling	Monitoring of community shifts; rapid comparative analysis	Identification requires additional 16S rRNA-based approaches

Although there is much to be desired, molecular biology technology and research strategy will be a main way to reveal the true level of diversity and species composition of the microbial community in oilfields. Its development will promote great progress in oil and environmental microbiology technology, and is of great practical significance.

ACKNOWLEDGMENTS

Dr. Shao Hong-Bo's laboratory is jointly supported by ChangJiang Scholar Innovation Team Projects of the Education Ministry of China, Innovation Team Projects of Northwest A&F University, the International Cooperative Partner Plan of the Chinese Academy of Sciences, the Cooperative & Instructive Foundation of State Key Laboratory of Soil Erosion and Dryland Farming on the Loess Plateau (10501-HZ), the Award Foundation of State Key Laboratory of Soil Erosion and Dryland Farming on the Loess Plateau, and the Open Foundation of State Key Laboratory of Soil Erosion and Dryland Farming on the Loess Plateau (10501-194).

REFERENCES

Bass, C. & Lappin-Scott H. (1997). The bad guys and the good guys in petroleum microbiology. *Oilfield Review*, Spring, 17~25.

Davidova, I., Hicks, M. S., Fedorak, P. M. & Suflita, J. M. (2001). The influence of nitrate on microbial processes in oil industry production waters. *J. Ind. Microbiol. 27*, 80–86.

Dennis, D. M. (1998). Microbial production stimulation [R]. Rock Mountain Oilfield Testing Center. March 31, FC970010.

Eckford, R. E. & Fedorak, P. M. (2002). Chemical and microbiological changes in laboratory incubations of nitrate amendment ''sour'' produced waters from three western Canadian oil fields. *J. Ind. Microbiol. 29*, 243 – 254.

Eckford, R. E. & P. M. Fedorak (2002). Planktonic nitrate-reducing bacteria and sulfate-reducing bacteria in some western Canadian oil field waters. *J. Indust. Microbiol. 29*, 83–92.

Ferris, M., Muyzer, G. & Ward, D. (1996). Denaturing gradient gel electrophoresis profiles of 16S rRNA-defined populations inhabiting a hot spring microbial mat community. *Appl. Environ. Microbiol, 62*, 340–346.

Gevertz, D., Telang, A. J., Voordouw, G. & Jenneman, G. E. (2000). Isolation and characterization of strains CVO and FWKO B, two novel nitrate-reducing, sulfide-oxidizing bacteria isolated from oil field brine. *Appl Environ Microbiol*, *66*, 2491–2501.

Gevertz, D., Telang, A. J., Voordouw, G. & Jenneman, G. E. (2000). Isolation and Characterization of Strains CVO and FWKO B, Two Novel Nitrate-Reducing, Sulfide-Oxidizing Bacteria Isolated from Oil Field Brine. *Appl. Environ. Microbiol.* p. 2491–2501.

Højris, B. & Skovhus, T. L. Enhancing the understanding of MIC through the application of molecular tools.

Hubert, C. & Voordouw, G. (2007). Oil Field Souring Control by Nitrate-Reducing Sulfurospirillum spp. That Outcompete Sulfate-Reducing Bacteria for Organic Electron Donors. *Appl. Environ. Microbiol.*, *73*, 2644–2652

Hubert, C., Nemati, M., Jenneman, G. & Voordouw, G. (2003). Containment of Biogenic Sulfide Production in Continuous Up-Flow Packed-Bed Bioreactors with Nitrate or Nitrite. *Biotechnol. Prog.*, *19*, 338-345.

Kim, J. & Kim, Y. (2001). *Corro. Sci.*, *43*, 2011.

Kleikemper, J., Schroth, M. H., Sigler, W. V., Schmucki, M., Bernasconi, S. M. & Zeyer, J. (2002). Activity and Diversity of Sulfate-Reducing Bacteria in a Petroleum Hydrocarbon-Contaminated Aquifer. *Appl. Environ. Microbiol, 68*, 1516–1523.

Kodama, Y. & Watanabe, K. (2003). Isolation and Characterization of a Sulfur-Oxidizing Chemolithotroph Growing on Crude Oil under Anaerobic Conditions. *Appl. Environ. Microbiol,* 69, 107–112.

Kuenen, J. G. (1989). Colorless sulfur bacteria. In: Nolt, J. G. (ed.)., *Bergey's Manual of Systematic Bacteriology*, Vol. 3. Williams and Wilkins, Baltimore, MO, pp. 1834– 1842.

Lane, R. A. (2005). Under the Microscope: Understanding, Detecting, and Preventing Microbiologically Influenced Corrosion. *Journal of Failure Analysis and Prevention, 5*, 10-12, 33-38.

Li, H., Yang, S. & Mu, B. (2007). Phylogenetic Diversity of the Archaeal Community in a Continental High-Temperature, Water-Flooded Petroleum Reservoir. *Curr Microbiol,* 55, 382–388.

Londry, K. L. & Suflita, J. M. (1999). Use of nitrate to control sulfide generation by sulfate-reducing bacteria associated with oily waste. *J. Ind. Microbiol,* 22, 582–589.

Lopez, I., Ruiz-Larrea, F., Cocolin, L., Orr, E., Phister, T., Marshall, M., VanderGheynst, J. & Mills, D.A. (2003). Design and Evaluation of PCR Primers for Analysis of Bacterial Populations in Wine by Denaturing Gradient Gel Electrophoresis. *Appl. Environ. Microbiol,* 69, 6801–6807.

Madsen, E. L. (2000). Nucleic-acid characterization of the identity and activity of subsurface microorganisms. *Hydrogeol J. 8*, 112– 125.

Murray, A. E., Hollibaugh, J. T. & Orrego, C. (1996). Phylogenetic compositions of bacterioplankton from two California estuaries compared by denaturing gradient gel electrophoresis of 16S rDNA fragments. *Appl. Environ. Microbiol,* 62, 2676–2680.

Muyzer, G. & Ramsing, N. B. (1995). Molecular methods to study the organization of microbial communities. *Water Sci. Technol.*, *32*, 1–9.

Muyzer, G., Dewaal, E. C. & Uitterlinden, A. G. (1993). Profiling of complex microbial populations by denaturing gradient gel electrophoresis analysis of polymerase chain reaction-amplified genes coding for 16s rRNA. *Appl. Environ. Microbiol, 59*, 695–700.

Myhr, S., Lillebø, BLP. & Sunde, E. (2002). Inhibition of microbial H_2S production in an oil reservoir model column by nitrate injection. *Appl. Microbiol. Biotechnol, 58*, 400–408.

Nemati1, M., Mazutinec, T. J., Jenneman, G. E. & Voordouw, G. (2001). Control of biogenic H2S production with nitrite and molybdate. *J. Ind. Microbiol.*, *26*, 350–355.

Nielsen, J. L. & Nielsen, P. H. (2002). Enumeration of acetate-consuming bacteria by micro-autoradiography under oxygen and nitrate respiring conditions in activated sludge. *Water Research, 36*, 421.

Nielsen, J. L. & Nielsen, P. H. (2002). Quantification of functional groups in activated sludge by microautoradiography. *Water Science and Technology. 46*, 389.

Orphan, V. J., Taylor, L. T., Hafenbradl, D., et al. (2000). Culture-Dependent and Culture-Independent Characterization of Microbial Assemblages Associated with High-Temperature Petroleum Reservoirs. *Appl. Environ. Microbiol,* p. 700–711.

Øvreas, L., Forney, L. & Daae, F. L. (1997). Distribution of Bacterioplankton in Meromictic Lake Sælenvannet, as Determined by Denaturing Gradient Gel Electrophoresis of PCR-Amplified Gene Fragments Coding for 16S rRNA. *Appl. Environ. Microbiol, 63*, 3367–3373.

Polz, M. F. & Cavanaugh, C. M. (1998). Bias in template to product ratios in multitemplate PCR. *Appl. Environ. Microbiol, 64*, 3724–3730.

Rolleke, S., Muyzer, G., Wawer, C., Wanner, G. & Lubitz, W. (1996). Identification of bacteria in a biodegraded wall painting by denaturing gradient gel electrophoresis of PCR-amplified gene fragments coding for 16S rRNA. *Appl. Environ. Microbiol,* 62, 2059–2065.

Sette, L. D., Simioni, K. C. M., Vasconcellos, S. P., Dussan, L. J., Neto, E. V. S. & Oliveira, V. M. (2007). Analysis of the composition of bacterial communities in oil reservoirs from a southern offshore Brazilian basin. *Antonie van Leeuwenhoek, 91*, 253–266.

Skovhus, T. L., Ramsing, N. B., Holmstrom, C., Kjelleberg, S. & Dahllof, I. (2004). Real-time quantitative PCR for assessment of abundance of Pseudoalteromonas species in marine samples. *Appl Environ Microbiol, 70*, p. 2273.

Sperl, P. L. & Sperl, G.T. New microorganisms and processes for MEOR. *Fossil Energy*. DOE/BC/14663-11.

Sublette, K. L., McInerney, M. J., Montgomery, A. D. & Bhupathiraju, V. (1994). Microbial oxidation of sulfides by Thiobacillus denitrificans for treatment of sour water and sour gases. In: Alpers, N. C. and Blowes, D. W. (eds.), *Environmental Geochemistry of Sulfide Oxidation*. American Chemical Society, Washington, DC, pp. 68–78.

Suzuki, M. T. & Giovannoni, S. J. (1996). Bias caused by template annealing in the amplification of mixtures of 16S rRNA genes by PCR. *Appl. Environ. Microbiol.*, 62, 625–630.

Teske, A., Wawer, C., Muyzer, G. & Ramsing, N. B. (1996). Distribution of sulfate-reducing bacteria in a stratified fjord (Mariager Fjord, Denmark) as evaluated by most-probable-number counts and denaturing gradient gel electrophoresis of PCR-amplified ribosomal DNA fragments. *Appl. Environ. Microbiol, 62*, 1405–1415.

Van Hamme, J. D., Singh, A. & Ward, O. P. (2003). Recent Advances in Petroleum Microbiology. *Microbiol. Mol. Biol. Rev., 67*, 503–549.

Videla, H. A. (1996). *Manual of biocorrosion*, Lewis Publishers. *Boca Raton*, pp. 37-67.

Von Wintzingerode, F., Gobel, U. B. & Stackebrandt, E. (1997). Determination of microbial diversity in environmental samples: pitfalls of PCR-based rRNA analysis. FEMS. *Microbiol. Rev.*, 21, 213–229.

Walker, J. J. & Pace, N. R. (2007). Endolithic Microbial Ecosystems . *Annu. Rev. Microbiol,* 61, 331–47.

Wang, J., Ma, T., Zhao, L., et al. (2008). Monitoring exogenous and indigenous bacteria by PCR-DGGE technology during the process of microbial enhanced oil recovery. *J. Ind. Microbiol. Biotechnol.*

Wang, W., Wang, J., Li, X., et al. (2004). Influence of biofilms growth on corrosion potential of metals immersed in seawater. *Materials and Corrosion/Werkstoffe und Korrosion,* 55, 30~35.

Watanabe, K., Kodama, Y. & Kaku, N. (2002). Diversity and abundance of bacteria in an underground oil-storage cavity. *BMC Microbiology*. 2:23doi:10.1186/1471-2180-2-23

Wawer, C., Ruggeberg, H., Meyer, G. & Muyzer, G. (1995). A simple and rapid electrophoresis method to detect sequence variation in PCR-amplified DNA fragments. *Nucleic Acids Res.*, 23, 4928–4929.

Yuan, S., Xue, Y., Gao, P., Wang, W. & Ma, Y. (2007). Microbial diversity in Shengli petroleum reservoirs analyzed by T-RFLP. *Acta Microbiologica Sinica,* 47, 290-294.

Zhang, W., Song, L., Ki, J., Lau, C., Li, X. & Qian, P. (2008). Microbial diversity in polluted harbor sediments II: Sulfate-reducing bacterial community assessment using terminal restriction fragment length polymorphism and clone library of dsrAB gene. *Estuarine, Coastal and Shelf Science,* 76, 682-691.

Zoetendal, E. G., Collier, L. S., Koike, K., et a1. (2004). Molecular ecological analysis of the gastrointestinal microbiota: A review. *J. Nutr.*, 134, 465-472.

Zumft, W. G. (1992). The denitrifying prokaryotes. In: Balows, A. B., Truper, H. G., Dworkin, M., Harder, V. & Schleifer, K. H. (eds.), *The Prokaryotes*, *Vol. 1.*, Springer-Verlag, New York, pp. 554–582.

In: Environmental Regulation: Evaluation, Compliance... ISBN: 978-1-60741-645-6
Ed: Diederik Meijer and Fillipus De Jong

Chapter 3

INCENTIVE-COMPATIBLE TARGETING FOR THE PROVISION OF PUBLIC GOODS IN AGRICULTURE

***D. Viaggi*[a*], *F. Bartolini*[a] *and M. Raggi*[b]**

[a] Department of Agricultural Economics and Engineering, University of Bologna, Viale Fanin, 50, 40127 Bologna, Italy

[b] Department of Statistics, University of Bologna, Via Belle Arti, 41, 40126 Bologna, Italy

ABSTRACT

Targeting is advocated as one of the main strategies to improve effectiveness of environmental policies. Among others, this is very relevant for EU agri-environmental schemes (AESs) through which payments are provided to farmers for the provision of environmental goods. Targeting requires support by way of appropriate knowledge systems and zoning, as well as the definition of priorities for funding when subsidies are involved. However, targeting often fails due to a lack of consideration of the economic incentives to participate in target areas. This is partly motivated by the fact that agents have private information on their compliance costs which is undisclosed to the regulator. The objective of this paper is to design a model of incentive compatible targeting strategies under asymmetric information. The model is then used to evaluate the impact of targeting strategies on designing agri-environmental contracts under adverse selection in different agricultural policy scenarios. The results suggest that targeting choices directly affect the optimal way contracts are designed and that this should be considered in the policy design process in order to avoid a failure of the targeting mechanisms. In the case analysed, different agricultural policy scenarios can significantly affect AES policy design.

Keywords: Agri-environmental schemes, contracts, multifunctional agriculture, agricultural policy reform

JEL classification: Q1 – Agriculture; Q18 - Agricultural Policy; Food Policy; Q2 - Renewable Resources and Conservation

* Corresponding author: davide.viaggi@unibo.it, tel. +39 051 2096114, fax +39 051 2096105

1. INTRODUCTION AND OBJECTIVES

The development of the multifunctional role of agriculture is accompanied by the diffusion of innovative forms of contracts aimed at creating incentives for the production of public goods by the sector. Some examples are the Agri-Environmental Schemes (AES) proposed under regulations 2078/92, 1257/99 and 1698/2005 whereby payments are provided to farmers for the provision of environmental goods. Given the increasing relevance of public services produced by agriculture, contract efficiency represents a key issue. In fact, the reduction of unjustified rents and the containment of policy transaction costs seem more and more necessary to guarantee the sufficient efficiency of public expenditure, the maintenance of the sector's credibility, and the ability to allow economic reward for those farmers actually producing public goods appreciated by society. Such needs are perceived at all levels, including the EU which has explicitly included environmental and social efficiency criteria in the mechanism of allocation of agri-environmental payments in light of the reform of rural development programs (reg. EC 1698/2005).

Targeting is advocated as one of the main strategies to improve environmental effectiveness of environmental policies. Among others, this is particularly relevant for AESs. Targeting requires support through appropriate knowledge systems and zoning as well as the definition of priorities for funding when subsidies are involved. However, targeting often fails due to the fact that economic incentives to participation are not considered. This can be motivated by the fact that agents have private information on their compliance costs which is undisclosed to the regulator.

The problem of contract efficiency has been dealt with by many authors through the lens of contract economics and, in particular, through principal-agent models. However, the problem of optimal contract design with targeting has barely been considered in the literature. This paper builds on the model previously developed in Bartolini et al., 2007 and explores the consequences of targeting choices for contract design.

The objective of this paper is to compare different ways of designing agri-environmental contracts under adverse selection when the interplay between contract design and targeting objectives is explicitly considered. The analysis is based on a principal-agent model under adverse selection, with explicit targeting constraints. The model is used for a numerical analysis carried out before and after the 2003 CAP reform, with the objective of analysing the interaction between the new CAP and the design of agri-environmental contracts. The case study and the examples given in the literature refer to Emilia-Romagna (Northern Italy).

The structure of this paper is the following. In section 2 the state of the art of AES analysis using the contract theory approach is illustrated together with the role of targeting. In section 3 the methodology is described followed, in section 4, by the illustration of the results of a case study. Some discussion is provided in section 5.

2. BACKGROUND

2.1. Scope of the Paper

The economic literature on contract design under asymmetric information has developed significantly in the last twenty years (Laffont and Tirole, 1993; Salanié, 1998; Laffont and Martimort, 2002).

Among the applications of this approach, AESs appear particularly pertinent. In fact, given that they are based on public goods production by numerous agents (farmers), diversified in terms of production costs (transaction costs included) and the degree of compliance, and being such costs/compliance farmer's private information, the contracts proposed by the public administration are most frequently designed based on partial knowledge of the relevant background information (primarily farmers' costs).

The problem is addressed in the literature in two different perspectives, the first being adverse selection, and the second moral hazard. The adverse selection problem is relevant due to the diversification of compliance costs among farmers, when the public decision maker is not able to distinguish between farmers belonging to different types. For its part, the moral hazard problem emerges when the public decision maker is unable to control the degree of contract compliance and incentives exist for the farmers to be totally or partially non-compliant.

In the following sections we deal with the argument in two parts. First, we discuss the relevance of this issue for the operational decisions concerning agriculture policy. Secondly, we briefly discuss the main contributions from agricultural economics literature. Finally, we address the issue of targeting and the connections with asymmetric information literature.

2.2. Relevance of the Problem

For its own characteristics, the assessment of the relevance of the issue of contract design under asymmetric information can hardly benefit from precise and unequivocal data. The data from mid-term evaluations of the rural development plans seem, however, to strengthen the idea that these problems are extremely relevant. For example, the economic impact evaluation of integrated production schemes in Emilia-Romagna shows that for peach and wheat production it would be profitable, on average, for farmers to participate in AESs, even without any payment. The same applies for organic wheat production (Emilia-Romagna Region, 2003a). This result may be due to the small differentiation in payments, with respect to the variety of farmers' compliance costs. The very high compliance cost differentiation among farmers is confirmed by results from monitoring carried out in Emilia Romagna. In fact, such monitoring highlights evidence of compliance costs ranging from less than zero to more than 500 euro/ha (unpublished data from Emilia Romagna Region, 2003a).

Clearly, if the results correspond to reality, the actual effectiveness of incentives would be very low, as they would have a minor effect on the cropping techniques. On the other hand, the results strongly support the expectation of adverse selection phenomena, whereby farmers who decide to participate are those with the lowest degree of change in farming practices in order to comply with the contract.

This evaluation, based on data reported by farmers, in most cases does not consider non-compliance. The degree of non-compliance is in fact very hard to measure. However, existing reports suggest that it could be very relevant. For example, the results from Emilia-Romagna controls show that only 57% of the farmers are totally compliant (Emilia-Romagna Region, 2003b). The remaining percentage may include only a small degree of non-compliance. On the other hand, most of the controls are based on some self-reporting, hence the degree of non compliance may be relevantly higher than detected.

The problem is relevant both for the design and for evaluation of AESs. In fact, even *ex post*, without objective indicators about the change in the state of the environment, evaluations based only on uptake (the most frequent case) would supply an illusory measurement of the environmental effectiveness and efficiency.

The argument applied so far about AESs takes on a more general importance in the light of the recent and projected evolution of the Common Agricultural Policy (CAP). In particular, the 2003 CAP reform introduced environmental cross-compliance on land benefiting from payment rights. In fact, the farmer is faced with the choice between benefiting from a payment with some restrictions and giving up the payment altogether with no obligations.

The experience from the USA for cross-compliance shows also the importance of tailoring farming constraints to compliance costs and the actual environmental problems of each area (Claassen et al., 2004). These issues, together with non-compliance, are a common feature of most cases of agri-environmental regulation (with and without payments) and represent a key factor affecting policy effectiveness and efficiency (OECD, 2004).

2.3. The Contribution of the Agricultural Economics Literature

The agricultural economic literature has tackled the problem since the beginning of the 1990s (Fraser, 1993) and a recent literature review on this topic may be found in Latacz-Lohmann (2004).

Briefly, the literature can be divided, with respect to the problems dealt with, into works that refer mainly to adverse selection, moral hazard or both.

The adverse selection problem is treated by Moxey et al. (1999) and Gren (2004), through models that hypothesise the possibility of providing farmers with a menu of contracts able to induce their self-identification through contract choice. These models are based on the maximization of social welfare; this function is composed of environmental improvement benefits, the benefit derived from the farmer's income increase and the cost of the distortionary effect of taxation, necessary to provide public funds. In the alternative, the problem may be posed in terms of cost-effectiveness, so eliminating the need to attribute monetary values to externalities and to quantify the distortionary effects of public funds (Havlik et al., 2003).

One alternative way of facing the problem is through contract auctions. This instrument was studied by Latacz-Lohmann and Van der Hamsvoort (1997; 1998) and has been addressed by several recent papers. On this topic, literature exists on the Conservation Reserve Program application in the United States, in which the contract assignments were based on auction mechanisms. In spite of the theoretical advantages of such mechanisms for contract allocation, practical experience shows a number of limits (on this issue see for example Viaggi and Taff, 2004).

The moral hazard problem is studied by different authors (for example Choe and Fraser, 1998; 1999). In general, the models used to deal with this problem attempt to identify the optimal monitoring level needed to ensure farmers' compliance. Policy design is represented through the optimal mix of payments, sanctions and the level of monitoring, in relation to farmers' compliance costs.

The contributions in which moral hazard and adverse selection are considered at the same time are less numerous (for example White, 2002; Ozanne and White, 2007). In this case, the main point is represented by the interaction (synergy) between incentives aimed at self-qualification and those aimed at compliance.

2.4. Targeting

An issue infrequently seen in the literature is the interplay between contract design and targeting objectives. Targeting is a key feature in policy design and is often used as a proxy for the social value of environmental improvements, assuming such value is correlated to the location of measures with respect to e.g. sensitive area or high natural value areas (OECD, 2004).

Increasing spatial targeting is considered to be one of the main areas of improvement and one of the main requirements for future efficiency of AESs (Latacz-Lohman and Hodge, 2003). Targeting may well be interpreted as an issue as long as it entails a trade-off between the precision of the AES and its cost, as it may be assumed that a higher targeting will imply a more than proportional increase in policy costs (Romstad, 2004).

Increasing targets may require different kinds of advances in policy. Some papers address the issue of targeting by proposing methodologies to qualify and select target areas (Khanna et al, 2003; Bayliss et al., 2003). While this is certainly a basic issue, it does not automatically show the policy design problem of giving proper incentives to farmers to ensure targeting of measures. A similar but more policy oriented problem is dealt with by papers discussing the characteristics of pollution measurement technologies, seen as policy support in devising effective targeting (Kurkalova et al., 2004a).

Targeting potential may be used as a discriminatory tool in selecting policy options (Duke and Lynch, 2007). However, this ability is often left to the intuition of evaluators. In other cases, different targeting strategies are explicitly used in simulating policy impacts, as in Kurkalova et al. (2004b), Ferraro (2004), Yang and Weersink (2004), Yang et al. (2005) and Watanabe et al. (2006). A common (and partially intuitive) result is that targeting may increase policy efficiency substantially when applied in an heterogeneous environment.

Most of these papers, however, do not directly consider the problem of targeting incentives when policy design is affected by information asymmetries. Fraser (2004) deals with this issue when analysing the problem of moral hazard in AESs. He finds that targeting may reduce policy costs, when costly monitoring and penalties are necessary to ensure compliance to policy descriptions. The paper also shows how monitoring/penalty parameters need to be adapted to the different target options in order to take advantage of targets in terms of enforcement costs. Targeting monitoring resources may be seen a strategic features to ensure the effectiveness of future AES policies (Fraser and Fraser, 2006).

This paper addresses a similar issue in the context of adverse selection. In particular, we analyse the implications of targeting objectives on the design of contracts using mechanism design.

3. The Model

In a previous paper, Bartolini et al. (2007) develop a principal-agent model under adverse selection, where the public regulator does not explicitly know the monetary value of the externalities produced by the sector. As a consequence, he has the aim of maximising the environmental improvement given the budget available. The type of contract assumed is limited to input (in particular nitrogen use) reduction[1]. It is assumed that the area of interest of the schemes has heterogeneous characteristics, as for both environmental sensitivity (and, as a consequence, priority in AES) and farm characteristics (compliance costs).

We develop this model further by assuming that it is in the interest of the regulator not only to maximise the environmental improvement, but also to pursue its concentration in priority areas. In order to do this, we design a model of optimal regulation and target-compatible incentives under three different conditions: a) perfect information (first best); b) asymmetric information with mechanism design (second best) and; c) flat payment, only differentiated between areas (which is the reference contract, i.e. the closest to reality).

The decision problem under perfect information conditions is modelled as follows:

$$\min z = \sum_{i=1}^{2} \sum_{j=1}^{m} \lambda_{ij} q_{ij} \tag{1}$$

s.t.

$$\text{bc: } \sum_{i=1}^{2} \sum_{j=1}^{m} \lambda_{ij} p_{ij} \leq B \tag{2}$$

$$\text{tc: } \frac{\sum_{j=1}^{m} \lambda_{1j} q_{1j}}{\sum_{j=1}^{m} \lambda_{1j}} \leq \tau \frac{\sum_{j=1}^{m} \lambda_{2j} q_{2j}}{\sum_{j=1}^{m} \lambda_{2j}} \tag{3}$$

$$\text{pc: } \pi_{ij}(q_{ij}) + p_{ij} \geq \pi_{ij}^{*}(q_{ij}^{*}) \tag{4}$$

where:

i = area type: 1=target, 2= non target;

j = farm type, j = (1, 2, …, m);

λ_{ij} = share of farms belonging to each combination of farm type and area type, with $\sum_i \sum_j \lambda_{ij} = 1$;

q_{ij} = maximum input quota to be assigned to each combination of farm type and area type;

p_{ij} = payment assigned to each combination of farm type and area type;

B = public budget available;

[1] The problem can be extended, in the same form, to others actions, characterised by input reduction (i.e. integrated production) and, with some variations, to the actions that result in a change in land use (i.e. wetlands).

τ = ratio between the nitrogen quota to be proposed in target areas and the quota to be proposed in non-target areas (degree of targeting).

$\pi_{ij}(q_{ij})$ = farm profit as a consequence of the quota assigned q;

$\pi_{ij}^{*}(q_{ij}^{*})$ = unconstrained profit.

bc is the budget constraint, limiting the amount of money available for the public to pursue its environmental objectives. The constraint *tc* forces solutions to satisfy the targeting objective τ. Mathematically, it guarantees a certain concentration of AES impacts (e.g. level of pollutant reduction) in target areas. This objective can be expressed in different ways. In the model adopted, the method chosen is that of a ratio between the maximum allowable use of input of target areas and that of non-target areas (roughly the same criteria applied in the nitrate directive).

The implementation of the regulator's program is constrained to the farmers' decision to participate (*pc*) This is actually differentiated according to the information conditions and to the form assumed for the payments. In the case of perfect information illustrated above (first best) it is assumed that the regulator knows the compliance costs of each individual farmer. Under this assumption, in order to guarantee participation at a minimum cost, it is sufficient for the public regulator to guarantee with equality that the payments are higher than the compliance costs (assuming reservation utility equal to zero), as in equation 4.

By maximising (1) constrained to (2), (3) and (4) we obtain the result, where the payment equals the compliance cost, calculated as the difference between the unconstrained and the constrained profit.

In the case of asymmetric information, equation (4) still holds, but the payment cannot be directly calculated on the basis of the compliance costs of each individual farmer, as they are not known to the regulator. However, the regulator can be assumed to know the value of compliance costs for each type of farmer, and have some prior expectation about the frequency of each farm type.

In this case, the best theoretical solution (revelation principle) is a menu of contracts built using mechanism design. The contracts are given by some combination of p and q for each farm type, such that:

$$\text{ic: } \pi_{ij}(q_{ij}) + p_{ij} \geq \pi_{ij}(q_{ij'}) + p_{ij'} \tag{5}$$

where:

j' = each farm type different from j.

The constraint is active only for farms in the same area given that, among areas, it is possible to discriminate according to farm location.

By maximising (1) constrained to (2), (3), (4) and (5), it is possible to obtain a different contract for each combination of farm and area type[2], each one designed in such a way that its selection is profitable for the farm for which it is designed.

One reference solution which is closer to the existing one is that of differentiating contracts between target and non-target areas, but not among different farm types. In this case

[2] This is not true in the case of pooling, that is not considered in this paper (see Laffont and Martimort, 2002).

the problem is solved by maximising (1) subject to (2), (3) and (4), with the payment and the quota indexed on *i* only (target or non-target area), and not on *j* (farm type).

4. A NUMERICAL EXAMPLE

4.1. The Case Study

The model has been applied to an illustrative case study, using data from the Municipality of Argenta (Ferrara, Emilia-Romagna). Two farm typologies were supposed, and distinguished on the basis of the different typology of technical-economic orientation: mainly cereal crops for farm 1 and mainly vegetables for farm 2.

Compliance cost functions may be regarded as dependent upon the policy framework in place. In this paper, cost functions have been estimated assuming two possible conditions: the first under the Agenda 2000 rules and the second under post-2003 CAP.

The gross margin (that substitutes the profit in this simulation) functions and the compliance cost functions, for the variation, respectively, of the quota on nitrogen use and the reduction of nitrogen use with respect to the private optimum are reported in Table 1.

Table 1. Income and compliance cost functions.

Farm Typology	Gross Margin Function	Compliance cost function
Type 1 Agenda 2000	$\pi(q) = 73{,}286+11{,}962\ q -0{,}0445\ q^2$	$c(r) = -0{,}3937\ r +0{,}0217\ r^2$
Type 1 2003 Reform	$\pi(q) = 286{,}9+9{,}1941\ q -0{,}0316\ q^2$	$c(r) = 3775{,}8\ r -18{,}715\ r^2$
Type 2 Agenda 2000	$\pi(q) = 7{,}8893+15{,}491\ q -0{,}0449\ q^2$	$c(r) = 1{,}0793\ r +0{,}0451 r^2$
Type 2 2003Reform	$\pi(q) = 84{,}816+13{,}727\ q -0{,}0366\ q^2$	$c(r) = 2866{,}3\ r -18{,}369\ r^2$

These functions are estimated using linear programming models, calibrated on structural data, derived to agricultural census, and technical data, derived to interviews to local experts (see more details in Bartolini et al., 2007). The gross margin and cost functions have been obtained by parametrising on a nitrogen use constraint (quota) and subsequently interpolating the response function obtained[3]. In order to carry out simulations, an homogeneous farm distribution has been hypothesised across the two typologies and across target and non target areas ($\lambda_{i,j}$ = 0,25 for all farm types/areas). The budget available to the regulator has been assumed to be equal to 300 euro/ha. Furthermore, we have hypothesised the objective to achieve a half use of nitrogen in target areas with respect to non-target areas.

[3] We do not include further details, since the single farm modelling does not represent the main objective of this paper.

4.2. Results

The optimal contract structure for the Agenda 2000 scenario shows clear differences between targeting and non-targeting options (Table 2).

Table 2. Results of targeting vs. Non-targeting Agenda 2000 (B=300 euro/ha, $\lambda_{i,j}$ = 0,25, target = 0,5).

		q (kg N/ha)				p (euro/ha)				average q (kg N/ha)
		Type 1		Type 2		Type 1		Type 2		
		t	nt	t	nt	t	nt	t	nt	
No target	First Best	51	51	90	90	303	303	297	297	71
	Second Best menu	42	42	119	119	483	483	117	117	81
	Second Best *q* uniform	90	90	90	90	300	300	300	300	90
Target	First Best	30	80	69	118	479	127	471	122	74
	Second Best menu	18	76	96	152	789	147	255	9	86
	Second Best *q* uniform	64	128	64	128	520	80	520	80	96

t=target area; nt=not target area

In both targeted and non-target conditions, the contract menu shows the ability, with the same budget, to propose a nitrogen use quota 10% less than a uniform area payment. Targeting shows some costs in terms of increase of farm quota (i.e. less overall reduction), but it is apparently not prohibitive in presence of strong motivations for concentrating participation in target areas (i.e. high value of the environmental good produced in target areas). However, the optimal contract structure (quota and payments by farm typology) results are totally diversified between targeting and non targeting options. As expected, quotas are stricter and payments higher in target areas. The impact is proportionally higher in the case of a menu of contracts (second best conditions), where the contracts must also provide incentive compatibility. For this reason, in this case, the quota is much lower for the more efficient type than in the other cases, and the payment much higher.

The simulation under the 2003 CAP reform conditions show a lowering of the quota for the first best and for the menu of contracts, while the quota increases (worsens) for the unified payment (Table 3).

This depends on the fact that the 2003 reform generally tends to induce a reduction of input use and so it has an effect in terms of higher ability to meet restrictions in input use.

However, it also increases the difference among opportunity costs in different farms, with respect to agri-environmental constraints, represented by the same nitrogen input quota. Therefore, a higher payment differentiation is needed in order to exploit the new situation in terms of optimal policy design. For the same reason, uniform contract solutions show a lower performance in this scenario.

Table 3.Results of targeting vs. Non-targeting – post-2003 reform (B=300 euro/ha, $\lambda_{i,j}$=0,25, target=0,5).

		q (kg N/ha)				p (euro/ha)				average q (kg N/ha)
		Type 1		Type 2		Type 1		Type 2		
		t	nt	t	nt	t	nt	t	nt	
No target	First Best	35	35	92	92	303	303	297	297	63
	Second Best menu	28	28	130	130	516	516	84	84	79
	Second Best q uniform	92	92	92	92	300	300	300	300	92
Target	First Best	14	61	74	114	464	142	436	158	66
	Second Best menu	1	64	109	156	809	205	186	0	83
	Second Best q uniform	65	130	65	130	514	86	514	86	97

t=target area; nt=not target area

As the differences among farm types become stronger, the difference between targeting and non-targeting also becomes stronger and this is more evident, once again, in the case of second best menu.

5. CONCLUSION

This paper adds further evidence to the literature on AES contract design, highlighting both the potential benefits from the adoption of menus of contracts as compared to flat rate payments, and the need to coordinate such menus with targeting strategies.

The paper also demonstrates the sensitivity of the results to policy changes. In particular, the 2003 CAP reform, through the decoupling process, is having an impact on AES compliance costs and hence tends to change optimal contract designs and the appropriate strategies for targeting AES measures.

The possibility and the strategy for contract improvement with respect to the flat rate option is a function, both of the variability of the cost for the production of environmental services among farms, and of the effective degree of asymmetric information among actors, also in relation to the growing investment for the collection and processing of farming data. In perspective, the higher complexity of the menu of contracts may be justified in some cases, while the use of fixed payments could remain the best solution in other cases.

The main issue addressed here, however, is the interplay between targeting and contract design. The example shows the relevance of targeting choice for contract design. On the one hand, targeting can be achieved at a relatively low cost in terms of reduction of environmental improvements (under budget constraints). On the other hand, trying to promote targeting through contract design implies major changes in contract design itself. This can be difficult to justify from the policy perspective and would certainly limit the practical ability to concentrate policy effects in target areas.

Different extensions and further research options can be identified for this paper. For example, the optimal contract structure could be studied in different hypotheses of correlation among priority areas and compliance costs. Moreover, transaction costs that are additional with respect to incentive and information costs could be taken into account. Finally, the externality value and the opportunity cost of public funds should be included for moving from the cost/effectiveness approach towards a more complete social cost and benefit analysis.

Acknowledgments

The work is based on reflections developed in the research "Instrument for economic exploitation of multifunctionality in agriculture", funded by the regional government of Emilia-Romagna, L.R. 28. Previous versions of some parts of this paper were presented at the XIth Congress of the EAAE (European Association of Agricultural Economists), Copenhagen, Denmark,: 23-27August, 2005.

References

Bartolini, F., Gallerani, V., Raggi, M. & Viaggi, D. (2007). Contract design and cost of measures to reduce nitrogen pollution from agriculture. *Environmental management, 40*, pp. 567-577.

Bayliss, J., Helyar, A., Lee, J. T. & Thompson, S. (2003). A multi-criteria targeting approach to neutral grassland conservation. *Journal of Environmental Management*, *67*, 145–160.

Choe, C. & Fraser, I. (1998). A note on imperfect monitoring of agri-environmental policy. *Journal of agricultural economics, 49*, 250-258.

Choe, C. & Fraser, I. (1999). Compliance monitoring and agri-environmental policy. *Journal of agricultural economics, 50*, 468-487.

Claassen, R., Breneman, V., Bucholtz, S., Cattaneo, A., Johansson, R. & Morehart, M. (2004). *Environmental compliance in US agricultural policy. Past performance and future potential*, Agricultural economic report 832, United State Department of Agriculture (USDA). Washington DC: USDA.

Duke, M. D. & Lynch, L. (2007). Gauging support for innovative farmland preservation techniques. *Policy Science, 40*, 123–155.

Emilia Romagna Region (2003a). *Relazione sullo stato di attuazione del piano regionale di sviluppo rurale 2000-2006 in Emilia Romagna. Annualità, 2002*. Bologna.

Emilia Romagna Region (2003b). *Rapporto di Valutazione intermedia al 2003*. Bologna.

Ferraro, P. J. (2004). Targeting conservation investments in heterogeneous landscape: A distance- function approach and application to watershed management. *American Journal of Agricultural Economics, 86(4)*, 905-918.

Fraser, I. & Fraser, R. (2006). Targeting monitoring resources to enhance the effectiveness of the CAP. *EuroChoices, 4 (3)*, 22-27.

Fraser, I. (1993). *Agri-environmental policy and discretionary incentive mechanism: the countryside stewardship scheme as a case study*. Manchester: The Manchester Metropolitan University. Department of Economics and Economic History.

Fraser, R. (2004). On the use of targeting to reduce moral hazard in agri-enviromental schemes. *Journal of Agricultural Economics, 55 (3)*, 525-540.

Gren, I.-M. (2004). Uniform or discriminating payments for environmental production on arable land under asymmetric information. *European Review of agricultural economics, 31*, 61-76.

Havlik, P., Jacquette, F. & Boisson, J. M. (2003). Agri-Environmental Agreements for Enhancing Biodiversity Production by Farmers in Bile Karpaty, Czech Republic: An Empirical Analysis of Agency Theory Application. *81st EAAE Seminar*. Copenhagen.

Khanna, M., Yang, W., Farnswoth, R. & Onal, H. (2003). Cost-effective targeting of land retirement to improve water quality with endogenous sediment deposition coefficients. *American Journal of Agricultural Economics, 85 (3)*, 538-553.

Kurkalova, L. A., Kling, C. L. Zhao, J. (2004a). Value of agricultural non-point source pollution measurement technology: assessment from a policy perspective. *Applied Economics, 36 (20)*, 2287-2298.

Kurkalova, L. A., Kling, C. L. & Zhao, J. (2004b). Multiple benefits of carbon-friendly agricultural practices: Empirical assessment of conservation tillage. *Environmental Management, 33 (4)*, 519-527.

Laffont, J. J. & Martimort, D. (2002). *The theory of incentives. The principal-agent model.* Princeton: Princeton University Press.

Laffont, J. J. & Tirole, J. (1993). *The theory of incentives in procurement and regulation.* Cambridge MA: MIT press.

Latacz-Lohman, U. & Hodge, I. (2003). European agri-environmental policy for the 21st century. *Australian Journal of Agricultural and Resource Economics, 47 (1)*, 123-139.

Latacz-Lohmann, U. (2004). Dealing with limited information in design and evaluating agri-environmental policy. *90th EAAE Seminar*, Rennes.

Latacz-Lohmann, U. & Van der Hamsvoort, C. (1998). Auctions as a means of creating a market for public goods from agriculture. *Journal of Agricultural Economics,* 49, 334-345.

Latacz-Lohmann, U. & Van der Hamsvoort, C. (1997). Auctioning conservation contracts: a theoretical analysis and an application. *American Journal of Agricultural Economics, 79*, 407-418.

Moxey, A., White, B. & Ozanne, A. (1999). Efficient contract design for agri-environmental policy. *Journal of agricultural economics, 50 (2),* 187-202.

OECD (2004). *Agriculture and the environment: lessons learned from a decade of OECD work*, Paris, OECD.

Ozanne, A. & White, B. (2007). Equivalence of Input Quotas and Input Charges under Asymmetric Information in Agri-environmental Schemes. *Journal of Agricultural Economics*, *58 (2)*, 260–268.

Romstad, A. E. (2004). Multifunctionality – focus on resource allocation. 90th EAAE Seminar, Rennes.

Salanié, B. (1998). *The economics of contracts. A primer.* Cambridge MA: MIT press.

Viaggi, D. & Taff, S. J. (2004). Public acquisition of property rights to serve agriculture/conservation policy: Lessons from Italy and the US, *9th Joint Conference on Food, Agriculture and the Environment*, Conegliano 30-31 August 2004.

Watanabe, M., Adams, R. M. & Wu, J. (2006). Economics of environmental management in a spatially heterogeneous river basin. *American Journal of Agricultural Economics, 88(3)*, 617–631.

White, B. (2002). Designing Agri-environmental policy with hidden information and hidden action: a note. *Journal of agricultural economics, 53*, 353-360.

Yang, W. & Weersink, A. (2004). Cost-effective targeting of riparian buffers. *Canadian Journal of Agricultural Economics, 52*, 17–34.

Yang, W., Sheng, C. & Voroney, P. (2005). Spatial targeting of conservation tillage to improve water quality and carbon retention benefits. *Canadian Journal of Agricultural Economics, 53*, 477–500.

In: Environmental Regulation: Evaluation, Compliance . . . ISBN: 978-1-60741-645-6
Ed: Diederik Meijer and Fillipus De Jong

Chapter 4

COMBATING CLIMATE CHANGE IN A SMALL OPEN ECONOMY: THE CASE OF AUSTRIA

Michael Getzner*

Department of Economics, Klagenfurt University (Austria), Universitaetsstrasse 65-67, 9020 Klagenfurt, Austria

ABSTRACT

Austria as a member state of the European Union (EU) committed itself to a reduction of Greenhouse Gas Emissions (GHG) by 13% (based on 1990 emissions) until the Kyoto protocol observation period of 2008-2012 while the European Union has the target to reduce GHG emissions overall by 8%. Austria as a small open economy may face several problems in achieving an ambitious reduction goal. However, there emerge a number of especially interesting advantages of pursuing strict environmental policies. The paper describes the current state of climate policies in Austria and gives an overview of environmental regulations both at the national and EU level. Austria has implemented the EU regulatory framework, e.g. by taking part in the emission allowance trading scheme, and by implementing national policies. However, efforts have not been sufficient since Austria did not fulfil the goals of reducing GHG emissions. Emissions increased by roughly 20% since 1990 instead of being reduced by 13%. The "gap" between actual and Kyoto emissions would have to be filled by strict environmental policies. To the contrary, Austrian politics does not place an emphasis on climate change policies but plans to spend over EUR 400 m on purchasing "hot air" at the international carbon markets in order to offset the increase in Austrian emissions until 2012. The paper discusses other policy options and their economic impacts. An overview shows that – besides purchasing emission allowances on international markets – domestic policies would lead to higher economic benefits. For instance, policy options include increasing energy efficiency and using more renewable energy sources. Special emphasis is laid on the specifics of strict environmental policies in small open economies as new environmental technologies – with Austria currently being a net-importer of such technologies – could increase the

* Corresponding author: Ph. +43 463 2700 4124, Fax +43 463 2700 99 4124, Cell +43 676 4129222, Email: Michael.Getzner@uni-klu.ac.at

international competitiveness of Austrian energy technologies on the international markets.

Keywords: Climate change policy, open economies, benefits of environmental policies
JEL Codes: Q5, H4

1. INTRODUCTION

The industrialized nations of Europe and the USA account for more than one third of global CO_2 and GHG (Greenhouse Gas) emissions, with annual per-capita emissions of 8 to 20 tons of CO_2 which are more than twice of Chinese per-capita emissions, and about 8 times higher than Indian per-capita emissions. The share of emissions, as well as per-capita emissions, are paramount for any policy aiming at stabilizing climate change. While the USA have been reluctant to join international efforts to reducing GHG emissions (e.g. delays in ratifying the Kyoto protocol; Bang et al., 2006), the European Union (EU) has passed several commitments to reduce emissions by increasing energy efficiency and the share of renewable energy sources (cf. European Commission, 2005 and 2007; Council of the European Union, 2008a; for details see section 3).

From an economic viewpoint, efforts to combating climate change have been pictured as providing several public goods, at the global level (e.g. reduction of global temperature increase) and at the regional/local level (e.g. reduction of local air pollution) (cf. Pittel and Rübbelke, 2008). While the characteristics of climate change as a global public good may lead to free-rider behavior, many single countries have taken individual efforts to stabilize or reduce GHG emissions. In fact, a mixture of national and international policies is most efficient in economic terms (cf. Hackl and Pruckner, 2003). However, EU policy makers have argued that European policies alone would harm the European economy while providing only a minor reduction to global GHG emissions. This argument is mirrored in the differentiation in the European system of tradable carbon emission certificates between industries facing global competitions vs. industries with less competition (Council of the European Union, 2008b). With respect to the Austrian economy, this argument may even be more pronounced and can be sketched in the following way. Austria's share of GHG emissions would be marginal; Austria's domestic energy-intensive industries are leading in reducing energy and carbon emissions (e.g. steel production). If Austria would agree to reduce energy and carbon emissions in the heavy industry sector to a larger extent, some industries had to shut down with the effect that industrial products such as steel would have to be imported from other (non-EU) countries. However, as Austria is leading in low-energy steel production, it would import steel from high-energy and high-carbon "pollution havens". Finally, jobs in the industrial sector would get lost, with an overall negative effect on the environment since energy use and carbon emissions would increase in total. As climate change is a global public good, Austria would hurt its economy and the global and local environment by trying to reduce carbon emissions in the steel sector even further (cf. BMWA, 2008).

Taking this background as the starting point for discussion, the current chapter presents information on GHG emissions of both the European Union and Austria in section 2, and gives an overview of existing international and national policies for combating climate change in section 3. Section 4 then presents a brief historical overview of climate change policies, and focuses on future policy options for Austria as a small open economy. While the European Union's policies provide tentative solutions to the global public good problem by policy coordination and compromise, there is still leeway for pro-active national policies even for small economies such as the Austrian. Section 5 discusses the results and concludes.

2. CO_2 EMISSIONS IN THE EUROPEAN UNION AND AUSTRIA

The 27 member countries of the European Union (EU27) are among the largest emitters of Greenhouse Gases (GHG). In EU27 countries, CO_2 emissions came up to 4.165 bn tons (1995) and increased to 4.269 bn tons (2005). GHG emissions (in CO_2 equivalents) amounted to 5.249 bn tons (1995) and to 5.177 bn tons (2005) in EU27 (Eurostat, 2008). The small growth of CO_2 emissions and the marginal reduction of total GHG emissions are due to the economic downturn of the new member states of Central and Eastern Europe after the fall of the Iron Curtain in 1989. Many of the energy-intensive industries were not competitive and broke down leading to a dramatic decrease in industrial production. In the industrialized countries of Western Europe (EU15), with the major exceptions of Germany and the United Kingdom, emissions grew from the early 1990s until today. Spain, Italy, and Austria were among the countries with the largest increases in CO_2 and GHG emissions. EU15 countries alone produce about 12% of global CO_2 emissions, about half of emissions of China (24%) and the USA (21%). With India (8%) and Russia (6%), these countries emit close to 70% of global GHG emissions. However, it is not only the total amount of emissions that is of great concern to policy makers, but also the high per-capita emissions in the industrialized countries. It has been put forward that per-capita emissions of 2 tons would be a tentative "sustainability threshold" that may stabilize the World's climate. US per-capita emissions of 19.6 tons per capita are roughly twice per-capita emissions of EU15 countries; China's and India's per-capita emissions are much lower, with 3.9 and 1.1 tons.

From a sectoral viewpoint, the energy sector (e.g. electricity production) and the burning of fossil fuels for heating buildings and providing warm water for households, is by far the largest emitter of GHG emissions (59%). The transport sector (e.g. cars and trucks, airplanes, trains) emits about 21% of GHG emissions. The EU27's agricultural sector accounts for 9% of emissions, and industrial processes emit about 8% of GHGs.

Austria's emissions are comparatively small with a share of roughly 1.8% of EU27's GHG emissions, given that Austria accounts for roughly 1.7% of EU population, and about 2.2% of EU GDP (Gross Domestic Product). However, Austria's per-capita CO_2 emissions of 7.9 tons are clearly above the sustainable per-capita emissions of 2 tons. Figure 1 shows that CO_2 emissions have constantly grown from about 35 m tons in 1960 to about 78 m tons in the last years. Since the early commitments to reduce CO_2 and GHG emissions (for details section 4) from the 1990s onwards, Austria's emissions grew by about 25%. Austria signed the Kyoto protocol with an ambitious target to reduce GHG emissions by 2008-2012 by 13%; currently, emissions are by more than 30% off the target. The main driving force of the growth of GHG emissions is economic growth, contrary to the widely discussed concept of the Environmental Kuznets Curve (EKC) (cf. Copeland und Taylor, 2004; Dasputa et al., 2002; Dinda, 2004). Additionally to the validity of the EKC hypothesis for Austrian CO_2/GHG emissions, Friedl and Getzner (2003), and Getzner (2009a), have tested whether the openness of the Austrian economy (65% of GDP are currently exported) and structural change (60% of GDP is produced in the service sector) contribute to a reduction of emissions. It turns out that economic growth still largely determines (fossil) energy use and GHG emissions, only slightly moderated by increasing imports and structural change.

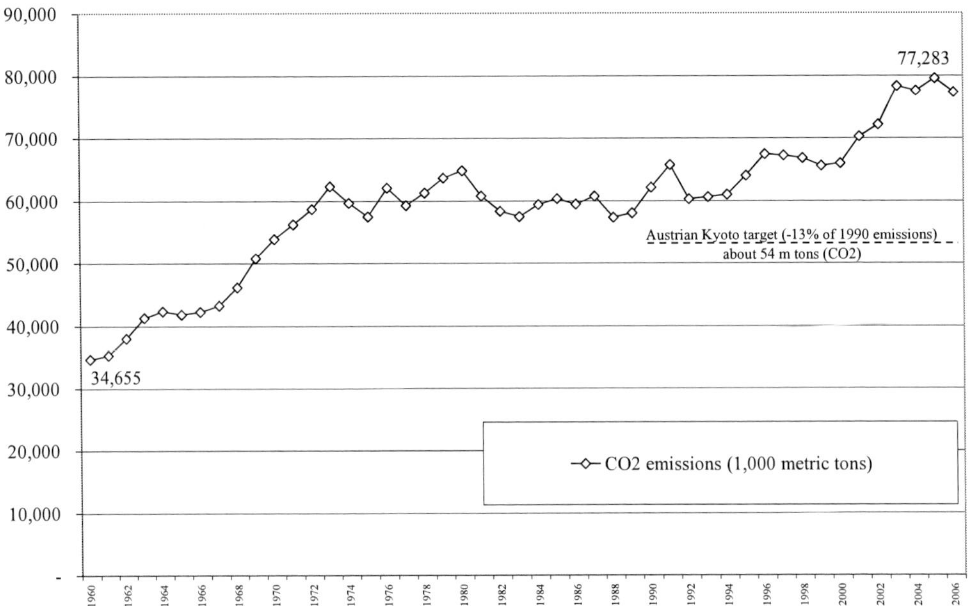

Figure 1. CO_2 emissions of the Austrian economy, 1960 to 2006 (1,000 metric tons)
Source: Umweltbundesamt (2008); author's calculations.

While emissions still grow in absolute figures, the energy and emissions intensity of the Austrian economy has decreased over time

Figure 2 shows that GDP growth was larger than the growth of GHG emissions; however, the energy efficiency of the Austrian economy has not improved substantially for about the last 10 to 15 years. A significant structural break can be detected in the mid-1970s following the oil price shock, with large efforts to reduce energy consumption. While the energy and household sector are by far the largest emitters, the transport sector with roughly 20% of total emissions contributes to the large growth of emissions. Getzner (2009b) presents evidence on the specifically large contribution of private car use to the growth of total CO_2 emissions. Passenger transports in private cars, and road freight transport, contribute significantly to Austrian CO_2 emissions, and especially to the growth of CO_2 emissions in the last 15 years.

3. Policies Implemented to Combat Climate Change

Combating climate change has been on the political agenda of the European Union for a long time. Only recently, the European Council decided at its December 2008 meeting to confirm EU's policies to reduce GHG emissions, to increase energy efficiency, and to increase the share of renewable energy sources. The energy and climate change strategies of the European Union are laid out in European Commission (2005 and 2007). The overall aim of EU's policies is to limit global warming to 2°C. As an international body, the EU is also part in international agreements, first of all, in the Kyoto protocol and the conferences and agreements according to the UNFCCC (2009). Recently, Streimikiene and Girdzijauskas (2009) have presented a detailed analysis of the Kyoto protocol and potential post-Kyoto regulations.

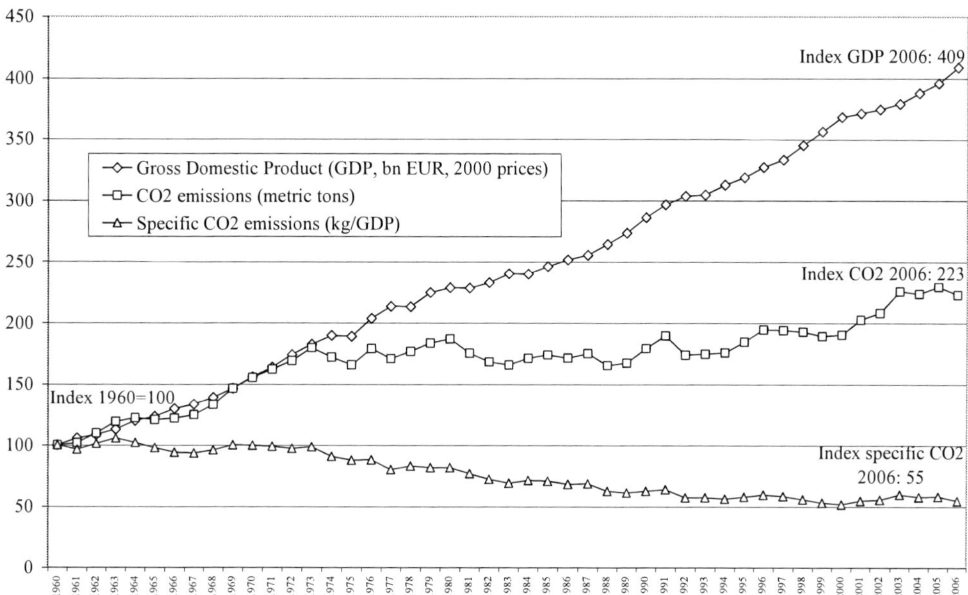

Figure 2. Index of CO_2 emissions of the Austrian economy, and GDP, 1960 to 2006 (1960=100)
Source: Umweltbundesamt (2008); Statistik Austria (2008); author's calculations.

Austria as a country and member of the European Union (since 1995) therefore takes part in the manifold international climate change negotiations. The probably most prominent and important international agreement for Austria is the Kyoto protocol. The European and national strategies mentioned above are by now unsuccessful in various degrees. While the EU15 target of reducing GHG emissions is –8% until 2008–2012, the EU15 countries managed to reduce CO2 emissions by only 2% until 2005. Ten EU27 member states increased their emissions while the remaining 17 member states managed to reduce their GHG emissions. Among those who contributed to EU's reduction in GHG emissions, Germany, Finland, the Netherlands and Romania did the most to combat climate change. For instance, Finland reduced its emissions by 14.6% while Germany reported a fall by 2.3%, the largest reduction in absolute terms (23.5 m tons). Spain recorded the largest absolute increase in GHGs, 15.4 m tons or +3.7%, while Austria falls short of its Kyoto reduction target by about 30%. Instead of reducing its emissions by 13%, emissions were about 20% above base year emissions. This development of CO_2 emissions in Austria has been severely criticized by several political stakeholders, including Austria's Government Accountability Office (GAO). In one of its recent reports, the office warns about the consequences of non-compliance

(Rechnungshof, 2008). The office estimates the financial costs for the Austrian government budget to amount to over EUR 0.5bn between 2008 and 2012 in order to compensate for surplus CO_2 emissions by means of the Kyoto institutions (Joint Implementation, JI; Clean Development Mechanism, CDM).

4. Austria's Policy Options as a Small Open Economy

Climate change policies have a rather long history in Austria's environmental policies. As early as the late 1980s, Austria has officially pursued a strict environmental policy including a number of commitments in policy documents, laws and regulations. For instance, one of the major arguments against the accession of Austria to the European Communities[1] (EC) in 1995 was that the environmental regulations existing at that time were stricter than in many other EC member countries and at the EC level. The Austrian umbrella organization of environmental NGOs put 200 environmental questions on the table to be answered by the Austrian government (BKA/WWF, 1993) stressing that many of the EC environmental regulations where softer than the Austrian ones. However, as Pesendorfer (2008) has recently summarized, Austrian environmental policies changed dramatically after 1995. After Austria's accession to the EC, the role of Austria as a first mover in environmental policies was changed in favor of economic and competition policies, at the cost of environmental policies. During the last 15 years, Austria's environmental policies only followed EC environmental regulations without major own initiatives.

While Austria's membership in the European Union (EU) forms the basis of about 80% of central government's policies fields such as economic, monetary and environmental policies, with a subsidiary or executive role of the national legislative and administrative institutions, Austria's climate change policies were much more avant-garde until 1995 (for the following, cf. Schewig and Zitz, 1992; Magistrat der Stadt Wien (MA 22), 1991; Österreichische CO_2-Kommission, 1994; Bundesministerium für wirtschaftliche Angelegenheiten, 1990 and 1993; Österreichische Bundesregierung [Bundesministerium für Umwelt], 1995). Austria was among the first nations to sign the Toronto Protocol (1988) following an international conference entitled "The Changing Atmosphere – Implications for Global Security". The Toronto Protocol obliged the signatory countries to reduce CO_2 emissions by 20% until 2005 based on 1988 emissions. Included in the protocol was an obligation for industrialized countries to higher control standards compared to developing countries. The Toronto Protocol, while included in the official environmental and energy policy documents of the early 1990s (Austrian Government's Energy Reports of 1990 and 1993; cf. Österreichische Bundesregierung, 1990 and 1993), formed the base line for climate change policies for several years. However, the impacts of this protocol were marginal since the diverse policies of the Austrian government mainly existed only on paper, without concrete, or even contradictory, policies such as transport policies (cf. e.g. Steininger et al., 2008).

In 1989, Austria signed the Noordwijk Agreement at the "Ministerial Conference on Atmospheric Pollution and Climatic Change" with the aim to stabilize CO_2 emissions until 2000 on the basis of 1989 emission levels. Another agreement (Bergen Agreement) was signed in 1990 at the "Conference on Sustainable Development in the ECR-Region" to stabilize emissions until 2000 based on 1990 emissions. Later in 1990, a number of industrialized (mainly European) countries including the European Communities signed a commitment at the Second World Climate Change Conference to stabilize emissions until 2000 based on 1988 emission levels.

In 1992, the so-called Climate Alliance was signed by several European cities and regions to aim at reducing CO_2 emissions until 2010 by 50% based on 1990 emissions. Austrian cities and regions have signed this agreement comprising about half of the Austrian population.

Austria was also signatory state of the UNFCC (United Nations Framework Convention on Climate Change) in 1992 (ratified 1994), and the subsequent international treaties, such as the Kioto Protocol.

This brief overview of historical development should highlight three major aspects of Austrian policies. First, Austria was definitely at the forefront of signing international agreements. The official government documents clearly refer to these commitments, in particular emphasizing the importance of concrete policies in the energy and transport sectors. Second, Austria's accession of the European Union led to a derogation of authority from the national to the EU level, with somehow comforting consequences for the Austrian policies. On the one hand, international (European) coordination of climate change policies led to a balancing of costs and benefits of such policies among European countries. On the other hand, many policies now decided on the EU level have been criticized by Austrian politicians, blaming the European Union for too strict environmental policies. Third, Austrian policies changed from a first-mover position to fulfilling the obligations of EU environmental regulations, with some of the regulations implemented insufficiently. For instance, Austria's implementation of EU's Natura 2000 regulations on establishing a network of protected areas for biodiversity conservation was subject to law suits before the Court of Justice of the European Communities (2007). It is often heard in Austrian politics that politicians would rather have different regulations that would benefit the Austrian economy or electorate to a larger extent. When politicians are reluctant or when they face intra-Austrian opposition, they are likely to blame "European legislation" or just the European capital "Brussels" to be responsible for legislation in order to shift the responsiblity for supporting or enforcing a rule to decision makers outside their "sphere of influence" (cf. Fallend, 2008). For instance, in climate change policies, efforts of the EU regarding the limitation of Greenhouse Gas Emissions (GHG emissions) was criticized by politicians as being too ambitious for Austria while the same politicians participated in the very negotiations regarding EU's climate change policy targets.

Regardless of the manifold treaties and agreements, Austria signed in the past, it is of importance how effective climate changes policies were. The governmental budget office of Austria, Rechnungshof (2008), has evaluated Austria's Climate Change Strategy 2002 and the adaptation of 2007 (Österreichische Bundesregierung, 2002; BMLFUW, 2007) in terms of its effectiveness. In many sectors of energy use, the strategy has proved ineffective, leading to a large gap between the political targets according to the Kyoto protocol (reduction of Austrian CO2 emissions by 13%) and the de-facto increase of emissions by roughly 20%.

---

[1] Nowadays, the European Communities have been incorporated in the European Union (EU).

In the context of policy options for Austrian climate change policies, a number of arguments and assumptions have to be made in order to draft a policy program. First, as Austria is a small open economy with roughly 8m residents, its share in global carbon emissions is marginal. While it is practically irrelevant for combating climate change whether or not Austria reduces its GHG emissions, it is nevertheless a political and moral obligation to contribute a fair share to GHG emission reductions, and to take responsibility for fossil fuel use. Recently, Anthoff et al. (2009) have discussed such issues in the context of marginal costs of climate change and the distribution implications associated with the unequal burden of climate change. Additionally, DeCanio (2009, 922) stresses that "no meaningful global carbon emissions reductions are going to be achieved unless the decision-makers of the great powers understand that the climate change problem is of real and vital interest to their nations. Reaching this conclusion will require guidance from the best available science, as well as acceptance of moral responsibility for the well-being of future generations."

Austria therefore has, on the grounds of ethics, a moral obligation to reduce its carbon emissions. However, reducing carbon emissions is also in the self-interest of the Austrian economy. Second, there is much evidence that strict environmental policies lead to substantial innovations which may increase international competitiveness on the world markets (Porter's hypothesis; cf. Porter and van der Linde, 1995). The Stern Review (Stern, 2007) stresses this argument by emphasizing the importance of technological change to combat climate change. For Austria, there are many policy fields in which reductions in GHG emissions in fact may lead to substantial economic effects. For instance, private households may profit significantly from saving energy in the fields of mobility, heating/cooling, and electricity consumption. Such policies may increase the available income of private households and therefore contribute to higher domestic consumption instead of importing fossil fuels.

Third, as Getzner and Nagy (2009) have pointed out, Austria's current imports of environmental technologies are about EUR 400m higher than environmental technology exports. Emphasizing the development and marketing of environmental technologies may therefore significantly contribute the Austrian Current Accounts in terms of export/import relations.

5. Summary and Conclusions

Numerous commitments to combat climate change have been ratified by the European Union as well by individual countries such as Austria. While some European countries have managed to reduce Greenhouse Gas (GHG) Emissions substantially, Austria has failed to reach the political targets of reducing GHG emissions. For instance, the Austrian targets of the Kyoto protocol to reduce GHG emissions by 13% have been failed leading to substantial needs to purchase carbon compensation at the international JI/CDM markets.

While there may many moral obligations to reduce GHG emissions exist (future generations, Intra-generational equity), it is also in the self-interest of even small open economies such as Austria to pursue climate-friendly policies. A number of advantages may be realized by implementing strict environmental policies. These advantages consist, on the one hand, of technological innovations in renewable energy production and use, and energy saving technologies. On the other hand, energy saving reduces expenses for fossil fuels, and therefore increases the available income of private households.

The compliance to national and international agreements was sought to be secured by the Austrian Climate Change Strategy. It turns out, that the measures and policies of the strategy did not have much effect. History and econometric evidence tells us that only the major price shock of the mid-1970 (oil price shock) increase energy efficiency substantially, and lead to a temporary stabilization of the level of fossil fuel use. For the future, it is open whether the international mechanisms for sanctions for non-compliance are strong enough to reduce carbon emissions. In the light of partial non-effectiveness of such agreements, new initiatives for a Europe-wide system of carbon taxes – that has already been discussed in the early 1990s – have to be taken.

References

Anthoff, D., Hepburn, C. & Tol, R. (2009). Equity weighting and the marginal damage costs of climate change. *Ecological Economics, 68(3)*, 836-849.

Bang, G., Froyn, C. B., Hovi, J. & Menz, F. C. (2006). The United States and international climate cooperation: International 'pull' versus domestic 'push'. *Energy Policy*, *35(2)*, 1282-1291.

BKA/WWF (1993). 200 Umweltfragen zum EG-Beitritt und Antworten der Bundesregierung. BKA (Bundeskanzleramt) and WWF (World Wind Fund for Nature), Vienna.

BMLFUW (2007). Anpassung der Klimastrategie Österreichs zur Erreichung des Kyoto-Ziels 2008-2013. Bundesministerium für Land- und Forstwirtschaft, Umwelt und Wasserwirtschaft (BMLFUW), Vienna.

BMWA (2008). Mitterlehner: "Energie- und Klimapaket ist Kompromiss im Sinne von Standort, Arbeitsplätzen und Klimaschutz". Press release of the Austrian Ministry for Economics and Labor, BMWA (Bundesministerium für Wirtschaft und Arbeit), 12 December 2008, Vienna.

Bundesministerium für wirtschaftliche Angelegenheiten (1990). Energiebericht der österreichischen Bundesregierung 1990. Wien.

Bundesministerium für wirtschaftliche Angelegenheiten (1993). Energiebericht der österreichischen Bundesregierung und Energiekonzept 1993. Wien.

Copeland, B. R. & Taylor, M. S. (2004). Trade, growth, and the environment. *Journal of Economic Literature, 42(1)*, 7-71.

Council of the European Union (2008a). Cover note from the Presidency to the Delegations regarding the Brussels European Council, 11 and 12 December 2008, Presidency Conclusions. Council of the European Union, 12 December 2008, 17271/08.

Council of the European Union (2008b). Note from the General Secretariat of the Council to the Delegations regarding Energy and Climate Change – Elements of the Final Compromise. Council of the European Union, 12 December 2008, 17215/08.

Court of Justice of the European Communities (2007). Judgment of the Court (Fourth Chamber) of 10 May 2007 – Commission of the European Communities v Republic of Austria (Case C-508/04): Failure of a Member State to fulfil obligations – Directive

92/43/EEC – Conservation of natural habitats and of wild fauna and flora – Measures transposing the directive), Luxemburg.

Dasputa, S., Laplante, B., Wang, H. & Wheeler, D. (2002). Confronting the environmental Kuznets curve. *Journal of Economic Perspectives*, *16 (1),* 147-168.

DeCanio, S. J. (2009). The Political Economy of Global Carbon Emissions Reductions. *Ecological Economics*, *69(3),* 915-924.

Dinda, S. (2004). Environmental Kuznets curve hypothesis: a survey. *Ecological Economics*, *49(4),* 431-455.

European Commission (2005). Winning the Battle against Global Climate Change. Communication from the Commission to the Council, the European Parliament, the European Economic and Social Committee, and the Committee of the Regions; Commission of the European Communities, 9 February 2005, COM (2005) 35 "final".

European Commission (2007). Limiting Global Climate Change to 2 degrees Celsius: The way ahead for 2020 and beyond. Communication from the Commission to the Council, the European Parliament, the European Economic and Social Committee, and the Committee of the Regions; Commission of the European Communities, 10 January 2007, COM (2007) 2 final.

Eurostat (2008). Europe in figures – Eurostat yearbook 2008. Office for Official Publications of the European Communities, Luxembourg.

Fallend, F. (2008). Euroscepticism in Austrian Political Parties: Ideologically Rootes or Strategically Motivated? In: Taggart, P., Szczerbiak, A. (eds), Opposing Europe? *The Comparative Party Politics of Euroscepticism.* Oxford University Press, New York, 201-220.

Friedl, B. & Getzner, M. (2003). Determinants of CO2 emissions in a small open economy. *Ecological Economics*, *45(1),* 133-148.

Getzner, M. (2009a). Determinants of (de-) materialization of an industrialized small open economy. *International Journal of Ecological Economics and Statistics, 14 (1),* 3-13.

Getzner, M. (2009b). Environmental impacts of personal mobility: exploring an Austrian EKC. In: Mazzanti, M., Montini, A. (eds.), Environmental Efficiency, Economic Performance and Environmental Policy. Routledge, London (forthcoming).

Getzner, M. & Nagy, M. (2009). Indikatoren-Bericht 2008 zu Wirtschaft/Umwelt. BMLFUW, Vienna.

Hackl, F. & Pruckner, G. (2003). How global is the solution to global warming? *Economic Modelling*, *20(1),* 93-117.

Magistrat der Stadt Wien, M. A. 22 (1991). Auskunft über das Klimabündnis und den diesbezüglichen Gemeinderatsbeschluß. Wien.

Österreichische Bundesregierung (2002). Strategie Österreichs zur Erreichung des Kyoto-Ziels (Klimaschutzstrategie 2002). Federal Chancellery, Vienna.

Österreichische Bundesregierung [Bundesministerium für Umwelt] (1995) (Hrsg.). Nationaler Umweltplan (NUP) Österreich. Bundesministerium für Umwelt, Wien.

Österreichische CO2-Kommission beim Bundesministerium für Umwelt, Jugend und Familie (1994). Empfehlungen 1993 der Österreichischen CO_2-Kommission (ACC) für ein Aktionsprogramm zur Erreichung des Toronto-Zieles. Akademie für Umwelt und Energie, Wien/Laxenburg.

Österreichische Raumordnungskonferenz (1992). Österreichisches Raumordnungskonzept 1991. ÖROK-Schriftenreihe Nr. 96, Wien.

Pesendorfer, D. (2008). Paradigmenwechsel in der Umweltpolitik – Von den Anfängen der Umweltzu einer Nachhaltigkeitspolitik: Modellfall Österreich? VS Verlag für Sozialwissenschaften, Wiesbaden.

Pittel, K. & Rübbelke, D. (2008). Climate policy and ancillary benefits: A survey and integration into the modelling of international negotiations on climate change. *Ecological Economics*, *68(2),* 210-220.

Porter, M. & Van der Linde, C. (1995). Towards a new conception of the environment-competitiveness relationship. *Journal of Economic Perspectives*, *9(4),* 97-118.

Rechnungshof (2008). Umsetzung der Klimastrategie Österreichs auf Ebene des Bundes, Report BUND 2008/11. Rechnungshof, Vienna.

Schewig, D. & Zitz, E. (1992). Lokale Maßnahmen zum Schutz globaler Klimaverhältnisse - entwicklungspolitische und ökologische Aspekte. Projekt im Auftrag der Steiermärkischen Landesregierung, erarbeitet vom Österreichischen Informationsdienst für Entwicklungspolitik (ÖIE) und vom Klimabündnis, Graz.

Steininger, K., Berdnik, S., Gebetsroither, B., Getzner, M., Hausberger, S. & Hochwald, J. (2007). Klimaschutz, Infrastruktur und Verkehr. Wegener Zentrum für Klima und Globalen Wandel, Karl-Franzens-Universität, Graz, Wissenschaftlicher Bericht 15.

Stern, N. (2007). The Economics of Climate Change. Cambridge University Press, Cambridge.

Streimikiene, D. & Girdzijauskas, S. (2009). Assessment of post-Kyoto Climate Change Mitigation Regimes Impact on Sustainable Development. *Renewable and Sustainable Energy Reviews*, *13(1*), 129-141.

UNFCCC (2009). United Nations Framework Convention on Climate Change. UNFCCC (www.unfccc.int, 7 January 2009), Bonn.

In: Environmental Regulation: Evaluation, Compliance... ISBN: 978-1-60741-645-6
Ed: Diederik Meijer and Fillipus De Jong

Chapter 5

BENEFITS OF COMPLIANCE WITH LOCAL ENVIRONMENTAL REGULATIONS: MNC PERSPECTIVES

Mai Anh Dao, George Ofori and Low Sui Pheng
National University of Singapore

ABSTRACT

The current study seeks to assess the perceived benefits of compliance with environmental regulations by multinational corporations (MNCs) operating in Vietnam. Scott's (2001) principle of "three pillars of institutions" is used to synthesize literature on compliance into a conceptual framework of the benefits of compliance with environmental laws and regulations. The study was based on face-to-face interviews with environmental managers of MNCs operating in Vietnam. The mean importance rating, t-test of the mean and factor analysis were used to test the validity of the benefits identified in the literature. It was found that firms had derived benefits from environmental compliance which were driven by cost-benefit calculations, moral values and social pressures.

INTRODUCTION

Since the early 1960s, the development of the environmental movement has placed more and more pressure on firms to acknowledge, characterize, analyze and report upon environmental issues and impacts. Many organisations and enterprises are aware that environmental issues are becoming urgent in all aspects of social life. In the effort to protect the environment, environmental management measures have been introduced on both a mandatory and a voluntary basis. Perceiving the benefits of compliance is crucial for firms to act in accordance with environmental laws and regulations.

The research seeks to assist government agencies in the development of effective environmental regulations, which necessitates the understanding of the motivations which undelie firms' compliance behavior. The overall objective of this chapter is, therefore, to assess the motivations of firms in their compliance behavior with regard to environmental laws.

Scott's (2001) principle of three pillars of institutions is adopted as the generic framework for the study. The framework presents an overarching model of institutions that helps to synthesize compliance literature across fields into a comprehensive model of compliance. Scott's (2001) three pillars of institutions group institutions under the regulative, normative and cognitive pillars. These pillars are used in this study as broad categories to delineate the benefits of compliance reviewed in the literature.

The hypotheses are tested and research questions answered using quantitative data from the survey. Mean importance rating and t-test of the mean are used as data analysis methods for this study.

The Conceptual Framework: Scott's Three Pillars of Institutions and Theories of Compliance

Theories of compliance provide distinct perspectives on what motivates compliance and noncompliance. This chapter reviews, organises and synthesises literature on compliance with regulations across the fields of management, psychology, sociology, and economics. The main points highlighted in the literature are categorized and synthesized into different groups of factors motivating firm compliance to environmental regulations.

Scott (2001), in his principle of three pillars of institutions, proposed a single coherent model for the study of institutions, which is employed as the generic theoretical framework for the synthesis of literature on motivations of compliance across fields into a conceptual framework of benefits of compliance. According to Scott, institutions are founded on three pillars: the regulative pillar, based on consequentiality; the normative pillar, based on appropriateness; and the social-cognitive pillar, based on orthodoxy. Benefits of firms' compliance behavior are developed around these three pillars of institutions, which are tested—and research questions regarding them answered—using quantitative data from the survey of the business community in Vietnam.

The Regulative Pillar

The regulative pillar represents a rational choice approach, viewing firms as choosing rationally among alternatives based on their calculations of expected consequences. In this regard, firms are seen as rational actors that act to maximize their economic self-interest. It is considered to be in the "actor's self interest to construct and maintain institutional structures that will govern not only others' but one's own behavior" (Scott, 1995; p. 67).

According to Becker's (1968) theory of rational crime, a profit-maximizing firm will comply with an environmental regulation only as long as the expected penalty of violating exceeds the compliance cost.

Spence (2001) studies the rational polluter model of the modern American environmental regulatory system, which is founded on the assumption that firms are rational and self-interested economic and political actors, and rational pursuit of their self-interest guides both their compliance decisions and their attempts to influence policy. In order to maximize profit, the rational polluter will shift as many costs as possible to society; one way it does so is by discharging its wastes into the environment. Even though the rational polluter may prefer a clean environment to a dirty one, it is individually rational for each polluter to continue to pollute. This is the lesson from Garrett Hardin's "tragedy of the commons" (Hardin, 1968), the prisoner's dilemma from game theory (Spence, 1995), Samuelson's (1954) analysis of public goods, Pigou's (1920) analysis of externalities, and other rational actor models of firm behavior. According to these views, rational polluters will pollute unless deterred by some sort of coercive action.

In studying the compliance behavior of taxpayers, Casey and Scholz (1991) extend the basic model of deterrence theory to focus on the cognitive processes and strategies people use in the decision-making process. The study identified several behavioral phenomena that are inconsistent with rational maximizing models of deterrence but that potentially affect compliance. It is suggested that taxpayers' decisions are sensitive to how risk information is presented and how preferences are expressed. When risks of noncompliance are known to the taxpayer, the preference reversal phenomenon suggests that the way preferences are expressed (e.g., whether a tax professional is used) can affect compliance decisions by altering the relative weight placed on the probability of detection versus the penalty if detected.

The regulative pillar refers to processes that constrain and regularize behavior: rule-setting, monitoring and sanctioning activities. The regulatory process in this sense concerns "the capacity to establish rules, inspect others' conformity to them, and, as necessary, manipulate sanctions—rewards or punishments—in an attempt to influence future behavior" (Scott, 2001; p. 52). Actors are said to conduct expedient behavior and force, and fear and expedience are considered to be the basis for compliance to an institution. The mechanism to ensure compliance, can, according to this pillar, be seen to be coercive. Rules and regulations are said to control these elements (coercion or expedience) of the regulative pillar. Laws, rules and sanctions can be seen as indicators of institutions in the context of this pillar and institutions are thought of as being legitimate because they are legally sanctioned. Scholars who emphasize the regulative element of institutions include economists (for example, Hardin, 1968; Samuelson, 1954) and rational political science theorists (for example, Riker, 1962).

The Normative Pillar

Normative theories follow the "logic of appropriateness," which sees actions as based on identities, obligations, and conceptions of appropriate action, or, as termed by some authors, moral acts or intrinsic motivation (see Tyler, 1990; Sutinen and Kuperan, 1999).

The heart of normative theories, in the context of this study, is that firms are institutions that are generally inclined towards compliance with environmental laws, whether because of civic motives, social motives, or internalization of societal norms favoring environmental protection. Firms are motivated to act in compliance with environmental laws and regulations because of their concern for social reputation (see, for example, Allingham and Sandmo, 1972). Social influence and morality are closely linked. The symmetry characteristic of moral acts implies that the standards used to judge one's own behavior are used to judge others' behavior. Therefore, the moral principles on which individuals base their own behavior are also the basis for the social influence they exercise. Social influence to conform is expected to be stronger the more widespread a common moral obligation is in the population.

Social influence plays a significant role in everyday social exchange, often taking the subtle forms of ostracism or withholding of favors. Like enforcement authorities, peer groups can reward and punish their members, either by withholding or conferring signs of group status and respect, or more directly by channeling material resources toward or away from a member of the group.

Community and peer groups are considered a source of influence on individuals' action. If peer groups are non-compliant, the individuals are likely to be non-compliant, too (Sutinen and Kuperan, 1999). As an example, it has been found that social influence in fisheries is often manifested in the form of verbal and physical abuse (for example, fist fights, and destruction of gear and vessels). In the Massachusetts lobster fishing industry, strong forms of social influence, commonly called "self-enforcement", are estimated to account for the bulk of enforcement in the fishery (Sutinen and Gauvin, 1988).

The Cultural-Cognitive Pillar

The cognitive dimensions of institutions mark the distinction of new institutionalism in sociology. The cultural cognitive elements present "the shared conceptions that constitute the nature of social reality and the frames through which meaning is made" (Scott, 2001; p. 57). The most important cognitive element is constitutive rules, which involve the creation of categories and the construction of typifications. "For cultural-cognitive theorists, compliance occurs in many circumstances because other types of behavior are inconceivable; routines are followed because they are taken for granted as "the way we do these things" (Scott, 2001; p. 57). The basis for compliance to an institution is also this "taking this for granted" and it is spread though mimicking others. Moreover, "a cultural-cognitive view stresses the legitimacy that comes from adopting a common frame of reference or definition of the situation" (Scott, 2001; p. 61).

MNCs and Their Environmental Performance in Vietnam

Since the early 1990s, Vietnam has achieved impressive economic development. The main force behind the country's economic growth is the large number of companies which have moved into Vietnam from foreign countries such as Japan, Singapore, Taiwan, and South Korea, Europe and the U.S., and an associated increase in the amount of direct investment.

By 2007, foreign enterprises had increased dramatically from previous years to account for 24.8% of the total investment in business activities in Vietnam (Table 1).

While the opening up of a market economy since 1989 has been successful in generating strong economic growth, it has created threats to the country's environment. The Vietnamese Government has coped with this situation by establishing environmental laws and regulations starting with the enactment of the Law on Environmental Protection (LEP) in 1994 together with other specific environmental regulations designed to deal with water pollution, air pollution, and industrial waste, which are the country's principal environmental challenges and, at the same time, the problems against which companies are required to take countermeasures.

Table 1. Number and proportion of enterprises by ownership, Vietnam 2003–2005

	2003	2004	2005	2006	2007
Number					
Total (billion VND)	239,246	290,927	343,135	404,712	521,700
State-owned enterprise	126,558	139,831	161,635	185,102	208,100
Non-state enterprise	74,388	109,754	130,398	154,006	184,300
Foreign investment enterprise	38,300	41,342	51,102	65,604	129,300
Proportion (%)					
State-owned enterprise	52.9	48.1	47.1	45.7	39.9
Non-state enterprise	31.1	37.7	38.0	38.1	35.3
Foreign investment enterprise	16.0	14.2	14.9	16.2	24.8

Source: *Vietnam Statistics Handbook 2008* (General Statistics Office of Vietnam)

Of significant effect on the operations of companies are the national standards called Vietnam Standards which give guidelines on water quality, air quality, and solid waste disposal. The important standards include Industrial Wastewater Discharge Standards (TCVN 5945-2005), Industrial Emission Standards-Inorganic Substances and Dusts (TCVN 5939-1995 and Industrial Emission Standards-Organic Substances (TCVN 5940-1995) specify the maximum allowable concentrations for 109 different hazardous chemical substances contained in emission gases. These standards need to be complied with. In practice, however, Vietnamese environmental administrative bodies are not enforcing these standards partly because there are too many substances which are subject to control, and because many of them are difficult to analyze (MOE (J), 2002).

Together with the development of environmental legislation, there has been growing concerns about environmental impacts of the operations of enterprises, pushing firms to apply pollution control measures. Foreign companies are among those with the highest environmental awareness within the business community in Vietnam (Dao, 2002). Firms that operate on a global basis are more concerned with environmental issues than Vietnamese ones. Joint ventures and companies owned outright by foreign investors (100% foreign invested companies) indicated a strong interest and have been applying various measures to protect the environment including conforming to ISO standards. In general, state-owned companies, while accounting for about half the nation's mining and manufacturing production, execute almost no environmental conservation measures.

For those domestic firms, environmental awareness and concepts such as the ISO 14000 Environmental Management System (EMS) are very new but there is a growing awareness that the EMS will be an important tool for prevention of pollution by industry. Multi-national companies appear to be more interested in implementing ISO 14000 EMS. Internal environmental concerns of other countries are being passed from international corporations down through their supplier networks in Vietnam (Dao, 2002). Automobiles, motorcycles, or electric appliances manufacturers, especially internationally well known brand names such as Honda, Yamaha and Toyota, have attracted much attention in Vietnam as well as in other countries for their environmental protection efforts. Companies that have financial and technological resources are expected not only to continue their steady environmental protection efforts but also to transfer technology and know-how related to environmental protection to local companies and to be a driving force for promoting environmental protection in Vietnam, progress in which is currently being impeded due to numerous problems.

Dao (2002) found that almost all companies in Vietnam have executed firm environmental conservation measures based on cost-benefit calculations including fear of fines and penalty. Those with high environmental awareness, on the other hand, implement conservation measures based on the principle that the environmental conservation measures constitute a normal corporate activity. This is partly because the parent companies of many of the firms are established enterprises that can promote similar environmental conservation measures to the extent possible wherever they operate, on the basis of their global environmental policies. This is also largely because their executives, mostly foreigners, have experienced environmental conservation measures in manufacturing plants in Japan. In addition to that, quite a few companies recognized reduction of energy cost and production cost through implementation of environmental conservation measures. Many of the MNCs Vietnam are internationally well known so their brand names are recognized as product names in Vietnam. For such companies, any environmental damage caused by them could harm their reputation and their brand images. This is one of the reasons why these companies are very keen to take environmental conservation measures (MOE (J), 2002; Dao, 2002).

By 2005, there were about 120 industrial estates and export processing zones in Vietnam. Some industrial estates, especially those managed by foreign investors such as Japanese and Singaporean corporations, although constituting only a small fraction of these establishments, exercise excellent environmental conservation measures, thereby contributing to the upgrading of the environmental conservation measures of Vietnam. These Japanese industrial estates naturally have their own environmental facilities such as wastewater treatment plants. In one industrial estate, the substances dealt with include some that are not listed in the Vietnamese standards, as its effluent standards are based on the Japanese experience of industrial pollution (MOE (J), 2002). The company managing this industrial estate requires the tenants to abide by the estate's standards including dealing with the 'additional' substances. It considers that preventing the industrial estate from causing environmental problems eventually leads to the protection of the interests of the tenant companies. A Japanese industrial estate even provides a termination clause in its tenant contract, in which the estate reserves the right to retire the tenant from the industrial estate if the tenant causes an environmental violation. The management company of the industrial estate first demands that the tenant causing an environmental violation should rectify the situation. If the tenant fails to rectify the situation the tenant has to leave the industrial estate. Tenants can enter this industrial estate only on condition that they will abide by this termination clause (MOE (J), 2002).

Currently, foreign-owned industrial estates tend to be mainly occupied by firms from the particular countries; in other words, Japanese and Singaporean companies operate in such as the Japanese and Singaporean owned and managed estates, respectively. However, there are some non-Japanese/Singaporean foreign companies operating in these industrial estates. It is expected that Vietnamese companies will also begin to operate in these industrial estates. In view of such a trend, the forward-looking environmental considerations by these foreign-owned industrial estates will greatly contribute to environmental conservation measures of Vietnam.

PRELIMINARY MODEL

Based on the framework of the three pillars of institutions, and the review of the literature on compliance theories, and on corporate environmental performance in Vietnam, a preliminary model of benefits of compliance to environmental laws as perceived by foreign firms operating in Vietnam is presented in Table 2.

Table 2. Priliminary model of firm compliance

Logic of action	Reasons for compliance (Attributes)
Regulative	
Consequences *Cost benefits calculations*	1. Avoid high non-compliance costs/sanctions
	2. Enable company to reduce material wastage
	3. Improve company's procedures
	4. Easy to integrate with other management systems
	5. Reduce company's operating costs
	6. Help to enhance company's productivity
Normative	
Appropriateness Identities, obligations, and conceptions of appropriate action	7. Improve workers' health, safety and welfare
	8. Enable company to contribute to efforts to protect the environment
	9. Considered essential in company's overseas drive
	10. Improve company's social reputation
	11. Increase company's competitiveness
Cognitive	
Orthodoxy Common beliefs Shared logics of action	12. Help to develop business culture of compliance with the law and environmental protection

RESEARCH METHODS

To test the hypothesized benefits of compliance with environmental laws as presented in Table 2, a questionnaire-based survey was chosen as the method of data collection to gather the views of firms regarding the matter under study. Thirty seven foreign companies responded to the request to participate in the survey. To meet the objectives of the study aiming at determining the motivations for compliance to environmental laws of foreign firms operating in Vietnam, the study targeted firms that had been certified to the ISO 14001 EMS as this tool is the most popular environmental management measure in Vietnam. Firms certified to this standard are considered to show high environmental awareness and thus, a good appreciation of the benefits of environmental management activities. The intent was to conduct the interviews primarily with environmental managers and environmental consultants involved in setting up ISO 14001 EMS for the company. Environmental managers were selected for several reasons: (i) it is envisaged that they would have a background in environmental issues; (ii) if a company has an environmental manager, or equivalent, then it would indicate that environmental issues are viewed as important in some way to the company; and (iii) the position of the environmental manager is generally at a senior level of management. For those organisations where an environmental manager was not available or such a position did not exist, a person in an equivalent position or a relevant senior manager within the company was interviewed.

In this study, to ensure the highest rate of response to the questionnaire, consdering the time and budget constraints, face-to-face interviews with the managers of the sample companies were undertaken. The field study was conducted over a three–week period in December 2007.

Mean importance rating and statistical test for difference between means were the main indicators used to analyse the survey data.

RESULTS AND DISCUSSIONS

Companies were asked to identify the motivations for compliance with environmental requirements and rank the identified reasons as follows: 1 = "not important" and 5 = "very important."

Mean importance ratings and t values for all the attributes regarding firm compliance with environmental laws are presented in Table 3.

Table 3. Ranking of motivations for firm compliance behavior to environmental laws

Reasons for compliance	Regulatory		
	Rank	Mean	T
Help to evelop business culture of compliance with the law and environmental protection	1	3.8367	7.097
Improve company's social reputation	2	3.6316	4.169
Avoid high noncompliance cost	3	3.5957	5.999
Increase company's competitiveness	4	3.5476	3.575
Improve workers' health, safety and welfare	5	3.4255	3.919
Considered essential in company's overseas drive	6	3.2895	1.924
Reduce company's operating costs	7	3.2558	1.425
Help to enhance company's productivity	8	3.0732	.464
Improve company's procedures	9	2.8889	-.927
Enable company to contribute to efforts to protect the environment	10	2.6667	-3.788
Enable company to reduce material wastage	11	2.6596	-2.183
Easy to integrate with other management systems	12	2.5814	-3.030

In implementing the regulatory environmental management measures, firms are most concerned with developing the business culture of compliance with the law and environmental protection, followed by protecting their reputation, avoiding sanctions for noncompliance, enhancing company's competitiveness, reducing operating cost, and facilitating access to the international market (t value larger than 1.645) (Table 3; Figure 1). The findings provide validation for both the rationalist and normative theories of compliance (Becker, 1968; Tyler, 1990; Scholz, 1998; Spence, 2001) and also Scott's view of the importance of all the three (regulative, normative and cognitive elements of institutions (Scott, 2001).

Foreign firms, which are usually large firms in Vietnam, seem to have a good perception of the economic benefits of environmental management of the firms' operations (i.e., reduced costs, reduced waste, improved working procedures). This is an issue that has been reflected in the literature; that larger firms are more likely to adopt environmental management plans in order to reduce costs (Henriques and Sadorsky, 1996).

The mean rating also shows foreign firms' high level of corporate responsibility reflected by the efforts in improving the working environment for the welfare of their workers. This reflects the business culture they have from overseas, which is still in the preliminary stage of development in Vietnam where there is a low level of awareness of environmental protection among the people and the local business community.

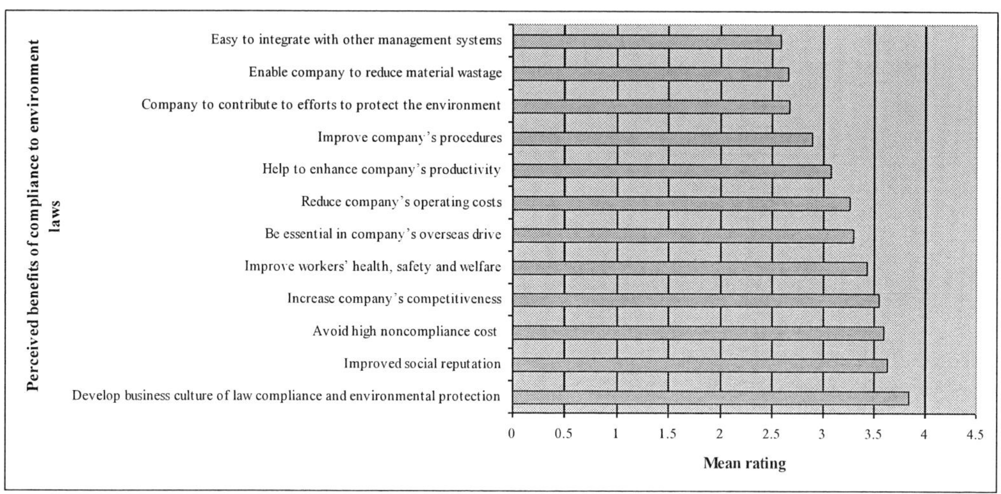

Figure 1. Ranking of benefits of compliance to environmental laws

The results of the t values and the mean importance ratings of attributes seem to indicate that the sampled firms were motivated to comply with environmental requirements by their awareness of compliance with the law, deterrence measures, social pressures and legitimacy of laws, and moral motives. This finding lends support to Scott's three pillars of institutions and theories of firm compliance where compliance behavior is considered to be based on a combination of the regulative, normative and cognitive elements of institutions. It also lends support to the indication that foreign businesses (majority of the sampled firms) in Vietnam are highly aware of environmental protection.

CONCLUSION

Understanding the factors which motivate the targetted entities to comply with regulations is key to the effectiveness of the law and regulations. Regulations governing aspects of the operations of businesses that can address the concerns of the businesses would achieve a high level of compliance. The investigation of firms' compliance behavior in the context of the Vietnamese business environment provides opportunities to develop a model of benefits of compliance that can be used to design appropriate policy-making strategies and instruments to enhance the environmental performance of businesses in the country.

The research findings show that MNCs, in the context of Vietnam, are motivated to comply with environmental regulations primarily because of their internalized norm of compliance. Developing a culture of compliance with the law and environmental protection within the firm is important for the businesses. Firms would also act in compliance with laws and show good environmental performance when they are under social pressure from customers, peer and interests groups or the general community. Morality is also important in determining firms' decisions to implement environmental management programs; as improving the firms' working environment and workers' health is ranked high.

Local companies in Vietnam would benefit from a programme to enhance their awareness of the need to minimize the adverse impact of their operations on the environment. Efforts could be made to educate them on the benefits of compliance with the laws in these regards. There is also scope for mentoring of the local firms by their MNC counterparts.

Additional research could involve comparative international studies so that the perceptions of businesses with respect to the benefits of complying with environmental regulations in other countries are investigated. The views of other businesses, including state-owned and private enterprises, could also be explored to broaden the picture of corporate performance in environmental management in Vietnam.

REFERENCES

Allingham, M. & Sandmo, A. (1972). Income tax evasion: a theoretical analysis. *Journal of Public Economics*, Vol. 1, pp. 323-338.

Becker, G. (1968). Crime and punishment: an economic approach. *Journal of Political Economy, Vol. 76, No. 2*, pp. 169-217.

Casey, J. T., & Scholz, J. T. (1991). Beyond deterrence: behavioral decision theory and tax compliance. *Law and Society Review, Vol. 25, No.* 4, pp. 821-844.

Dao, M. A. (2002). ISO 14001 Certification and Implementation in Vietnam. Unpublished M.Sc (Environmental Management) Dissertation, National University of Singapore.

Hardin. G. (1968). The tragedy of the commons. *Science, Vol. 162*, pp. 1243-1248.

Ministry of Environment (MOE (J)), Japan (2002) Overseas Environmental Measures of Japanese Companies (Vietnam).

Henriques, I. & Sadorsky, P. (1996). The Determinants of an environmentally responsive firm: an empirical approach. *Journal of Environmental Economics and Management, Vol. 30*, pp. 381-384.

Pigou, A. C. (1920). *The Economics of Welfare*. Cosimo Classics 2006.

Riker, W.H. (19620. *The Theory of Political Coalitions*. New Haven: Yale University Press.

Samuelson, P. A. (1954). The pure theory of public expenditure. *Review of Economics and Statistics*, *Vol. 36*, pp. 387-389.

Scholz, T. J. (1998). Enforcement policy and corporate misconduct: the changing perspective of deterrence theory. *Law and Contemporary Problems*, *Vol. 60*, *No. 3*, pp. 253-268.

Scott, W. R. (1995). *Institutions and Organisations.* CA, Thousand Oaks: Sage Publication.

Scott, W. R. (2001). (2nd Edition) *Institutions and Organisations.* CA, Thousand Oaks: Sage Publication.

Spence, D. B. (1995). Paradox lost: logic, morality, and the foundations of environmental law in the 21st century. *Columbia Journal of Environmental Law*, *Vol. 20*, pp. 145-128.

Spence, D. B. (2001). The shadow of the rational polluter: rethinking the role of rational actor models in environmental law. *California Law Review*, *Vol. 89*, *No. 4*, pp. 917-998.

Sutinen, J. G. & Gauvin, J. R. (1988). Enforcement and compliance in the commercial inshore lobster fishery of Massachusetts. *A report to the Environmental Enforcement Division, State of Massachusetts*.

Sutinen, J. G. & Kuperan, K. (1999). A Socio-economic theory of regulatory compliance. *International Journal of Social Economics*, *Vol. 26*, No. *1*, *2*, *3*, pp. 174-193.

Tyler, T. R. (1990). *Why Do People Obey the Law*. Yale University Press, New Haven.

In: Environmental Regulation: Evaluation, Compliance . . . ISBN: 978-1-60741-645-6
Ed: Diederik Meijer and Fillipus De Jong

Chapter 6

ULTRAVIOLET-INDUCED DAMAGE IN THE SKIN AND CORNEA: IMPLICATION FOR INFLAMMATORY CYTOKINE, MACROPHAGE MIGRATION INHIBITORY FACTOR

Tadamichi Shimizu[1*]
[1]Department of Dermatology, Graduate School of Medicine and Pharmaceutical Sciences, University of Toyama, Sugitani, Toyama, Japan.

ABSTRACT

Ultraviolet (UV) irradiation represents a significant environmental and occupational hazard that can cause acute and chronic inflammatory changes in the exposed skin and cornea. The inflammatory changes of acute exposure include erythema (sunburn) of the skin and photokeratitis of the cornea. Chronic exposure to solar UV irradiation leads to photoaging, immunosuppression and ultimately carcinogenesis in the skin. After skin and cornea damage by UV radiation, these tissues are known to secrete a number of cytokines, including interleukin (IL)-1, IL-6 and tumor necrosis factor (TNF)-α. Macrophage migration inhibitory factor (MIF) was originally identified as a lymphokine that concentrates macrophages at inflammatory loci, and it is a potent activator of macrophages *in vivo* which is considered to play an important role in cell-mediated immunity. Since the molecular cloning of MIF cDNA, MIF has been re-evaluated as a proinflammatory cytokine and pituitary derived hormone that potentiates endotoxemia. MIF is ubiquitously expressed in various tissues, including the skin and cornea. This article reviews the latest findings on the roles of MIF with regard to UV-induced damage in the skin and cornea.

Key words: skin, cornea, cytokine, macrophage migration inhibitory factor, inflammation, photoaging, photokeratitis.

---

* Corresponding author: Department of Dermatology, Graduate School of Medicine and Pharmaceutical Sciences, University of Toyama, Toyama, Japan.Tel: 81-76-434-7305 Fax: 81-76-434-5028 e-Mail: shimizut@med.u-toyama.ac.jp

INTRODUCTION

The effects of sunlight have fascinated researchers for decades because nearly every living organism on earth is likely to be exposed to sunlight, including its ultraviolet (UV) fraction it. UV is divided into three subtypes, each of which has distinct biological effects: UVA (320- 400 nm), UVB (280- 320 nm) and UVC (200- 280 nm), Although UVC is blocked by stratospheric ozone, UVB (1-10%) and UVA (90-99%) reach the surface of the earth and cause skin and eye damage [1]. Recently, the increased danger for the skin and eye from UV light has been linked to the decrease in stratospheric ozone. The health risks associated with ozone depletion are caused by enhanced UVA irradiation in the environment and increased penetration of UVB light [2, 3]. The skin is the body's main interface with the environment and frequently exposed to UV light. UV irradiation substantially increases the risk of actinic damage to the skin. UV irradiation may trigger cutaneous inflammatory responses by stimulating epidermal keratinocytes and fibroblasts to produce biologically potent cytokines such as interleukin (IL)-1 [4,5], IL-6 [6,7] and tumor necrosis factor (TNF)-α [8,9]. These cytokines are not only involved in the mediation of local inflammatory reactions within the skin, but they also exert systemic effects through entrance into the circulation. The eye is also exposed to UV irradiation. Acute exposure of artificial sources such as tanning lamps can result in severe pain and inflammation in the cornea. In addition, since the main UV source in nature is the sun, the solar UV dose received on a sunny day while skiing or sailing often causes acute photokeratitis when the eyes are unprotected [10].

MACROPHAGE MIGRATION INHIBITORY FACTOR (MIF)

Macrophage migration inhibitory factor (MIF) was originally identified as a lymphokine that concentrates macrophages at inflammatory loci. It is a potent activator of macrophages *in vivo* and it is also reported to play an important role in cell-mediated immunity [11, 12]. Since the molecular cloning of MIF cDNA [13], MIF has been re-evaluated as a proinflammatory cytokine and pituitary derived hormone that potentiates endotoxemia [14, 15]. Subsequent work has shown that T cells and macrophages secrete MIF in response to glucocorticoids as well as upon activation by various pro-inflammatory stimuli [16].

MIF is primarily expressed by T cells and macrophages; however, recent studies have revealed that this protein is ubiquitously expressed by a variety of cells, thus indicating a more far-reaching non-immunological involvement in a variety of pathologic states [17, 18, 19, 20]. In the skin, MIF is expressed in the epidermal keratinocytes and fibroblasts [21, 22]. MIF is known to play an important role in the skin in immune responses, inflammation and cell proliferation [23, 24, 25]. MIF is also detected in the cornea, iris, ciliary body and retina in the eye [26, 27] and the MIF expression increased when the cornea healed from surgical wound in rats [28]. It is thought that the MIF expression at the corneal basal membrane may mediate cell growth and differentiation [26, 29].

ULTRAVIOLET RADIATION AND THE SKIN; ROLE OF MIF

The inflammatory changes of acute exposure of the skin include erythema (sunburn), the production of inflammatory mediators, the alteration of vascular responses and the presence of inflammatory cell infiltrate into the skin. An enhanced MIF production in the skin has been reported after UV irradiation [30]. Solar UV light reaching earth is a combination of both UVB and UVA wavelengths, each of which stimulate MIF production in both keratinocytes and fibroblasts in the skin.

The detrimental long-term UV effect is cutaneous photoaging. Photoaged skin is biochemically characterized by a predominance of abnormal elastic fibers in the dermis and by a dramatic decrease in certain collagen types. Interstitial collagens, the major structural component of the dermis, are particularly reduced in UV irradiated actinically damaged skin [31, 32, 33]. There are several morphological and biochemical indicators that the amount and normal structure of collagen type I is reduced in UV actinically damaged skin [34]. Various types of UV-induced matrix-degenerating metalloproteinases, which are present in dermal fibroblasts, have been reported to contribute to the breakdown of dermal interstitial collagen and other connective tissue components. The underlying biological mechanisms causing this skin damage, involve a number of secreted cytokines, including interleukin (IL)-1, IL-6 and TNF-α [4, 6, 8]. UV irradiation up-regulates the production of these cytokines and the production of UV-induced collagenases such as matrix metalloproteinase (MMP)-1 from dermal fibroblasts is mediated in part by an IL-1β autocrine mechanism [35]. UVA stimulation leads to a significant increase in MIF mRNA and protein levels in human dermal fibroblasts [22]. MIF up-regulates MMP-1 mRNA and MMP-1 activity in dermal fibroblasts. Furthermore, MIF is involved in the up-regulation of UVA-induced MMP-1 in dermal fibroblasts through PKC-, PKA-, src-family tyrosine kinase-, MAPK-, c-jun- and AP-1-dependent pathways [22]. IL-1β stimulation leads to a significant increase in specific MIF mRNA and protein levels in human dermal fibroblasts [36]. In addition, neutralizing anti-MIF antibody suppresses the expression of MMP-1 induced by IL-1β. On the other hand, MIF is unable to stimulate IL-1β expression and production in dermal fibroblasts. It is therefore possible that after UVA irradiation, IL-1β induce MMP-1 expression. In addition, MIF induced by IL-1β also enhance the MMP-1 expression in dermal fibroblasts. In this context, dermal fibroblasts from MIF-deficient mice were much less sensitive to IL-1β-induced MMP-13 production (MMP-13 plays a restricted role in human tissues, it is the predominant tissue collagenase in rodents) [36]. It is therefore possible that IL-1β may stimulate MMP-13 production via MIF in dermal fibroblasts. In fibroblasts, UVA irradiation may stimulate IL-1β production by an autocrine loop of both IL-1β and MIF. Thereafter, both IL-1β and MIF play an important role in the synthesis of MMP-1 (MMP-13) (Figure 1).

UVA irradiation stimulates the production of granzyme B (serine protease) through MIF [37]. UVA induced granzyme B was able to degrade dermal fibronectin. Therefore, MIF appears to be an important mediator for the production of matrix-degenerating proteases, at least MMPs and granzyme B after UV exposure.

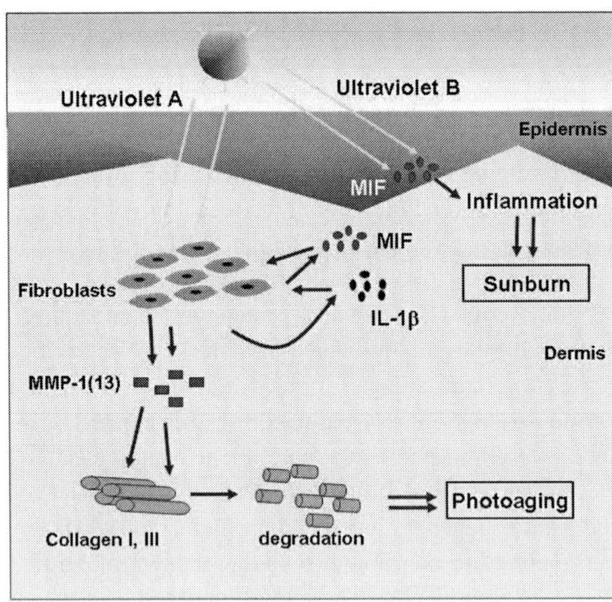

Figure 1. Schematic view of MIF related skin damage. Acute UVB light leads to an increased secretion of MIF by keratinocytes, which induces skin inflammation (sunburn). Chronic UVA irradiation stimulates MIF production in dermal fibroblasts by an autocrine loop involving both MIF and IL-1β. MIF mediates the up-regulation of M zMP-1(13), which thus contributes to the breakdown of dermal interstitial collagen.

ULTRAVIOLET RADIATION AND THE EYE; ROLE OF MIF

The eye is exposed to high levels of radiation of the optical spectrum and most of the ultraviolet and infrared radiation is absorbed in the cornea and lens. The recent enhancement of UV on the surface of the earth due to stratospheric ozone depletion may increase the risk of photochemically induced ocular damage [38]. The segment damage of the anterior eye from UV exposure includes effects on the cornea, lens, iris and associated epithelial and conjunctival tissues [39, 40, 41]. Acute UVB exposure causes corneal edema and photokeratitis and is associated with an inflammatory reaction in the cornea [42]. Previous studies have shown that the damage goes deeper than the epithelium, in fact, the response involves the full corneal thickness [43, 44, 45]. UV induces apoptosis in corneal cells [46] and recent studies described the influence of UVB on the transduction pathway of corneal cells; UV-induced apoptosis of corneal epithelial cells associated with K^+ channel activity [47, 48] and produced matrix metalloproteinase (MMP)-2 from corneal fibroblasts after UVB exposure [49]. MIF is upregulated by UVB irradiation in the mouse cornea [50] and MIF-overexpressed transgenic mice have milder corneal damages, fewer apoptotic cells and more c-jun positive cells than wild-type mice in response to UV radiation (Figure 2). MIF overexpression was originally thought to worsen the damage from UV radiation. However, that may be too simplistic a view to be true in this model. MIF may have cell protective functions in the eye. MIF expression may thus be related to the amelioration of UVB-caused corneal injury and this association is attributable to the upregulation of cell proliferation after acute UV-induced corneal damage, which involves the c-Jun dependent pathway [50]

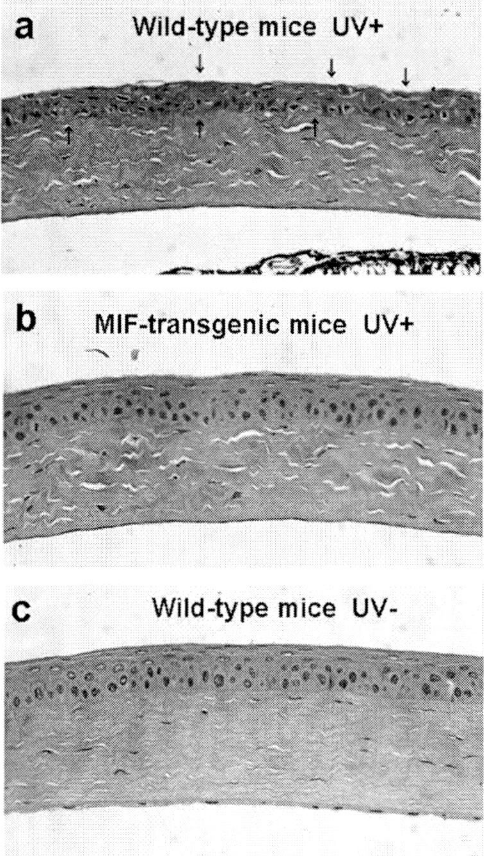

Figure 2. UVB irradiated cornea in MIF-overexpressed transgenic and wild-type mice. **(a)** Twenty-four hours after 400mJ/cm^2 UVB irradiation, the corneal basement membrane was destroyed and a corneal epithelial defect and thinning of the epithelium were observed in wild-type C57BL/6 mice irradiated by UVB. The destroyed corneal basement membrane and surface layer of the epithelium are indicated by arrows. **(b)** On the contrary, the corneal basement membrane of the MIF-transgenic mice was well preserved and the thickness of epithelium looked nearly normal. **(c)** Non-irradiated wild-type mouse cornea is shown as control. (hematoxylin&eosin staining).

Conclusion

Solar UV light reaching the earth is a combination of both UVB and UVA wavelengths, each of which stimulates MIF production in both skin and corneal cells. Although it has long been speculated that proinflammatory cytokines may play an important role in skin and ocular tissue damages, the mechanism has never been fully elucidated. Previous studies have demonstrated that the UV-induced cytokine network consisting of IL-1β and MIF interrelated loops induce MMP-1 and thus may contribute to the loss of interstitial collagen in cutaneous photoaging. Furthermore, the cornea is less damaged and can make a quick recovery when ocular tissue is sufficiently supplied with MIF. This newly identified mechanism may therefore help to elucidate the mechanism of photo-induced skin and ocular damage.

Acknowledgments

These experiments were carried out in collaboration with researchers whose work appears in the References.

References

[1] Matsumura, Y; Ananthaswamy HN. Short-term and long-term cellular and molecular events following UV irradiation of skin: implications for molecular medicine. *Expert Rev Mol Med.*, 2002, 4, 1-22.

[2] Tenkate, TD. Occupational exposure to ultraviolet radiation: a health risk assessment. *Rev Environ Health.*, 1999, 14, 187-209.

[3] Ma, W; wlaschek, M; Tantcheva-Poór, I; Schneider, LA; Naderi, L; Razi-Wolf, Z; Schüller, J; Scharffetter-Kochanek K. Chronological ageing and photoageing of the fibroblasts and the dermal connective tissue. *Clin Exp Dermatol,* 2001, 26, 592-9.

[4] Kupper, TS; Chua, AO; Flood, P; McGuire, J; Gubler, U. Interleukin 1 gene expression in cultured human keratinocytes is augmented by ultraviolet irradiation. *J Clin Invest,* 1987, 80, 430-6.

[5] Rutter, JL; Benbow, U; Coon, CI; Brinckerhoff, CE. Cell-type specific regulation of human interstitial collagenase-1 gene expression by interleukin-1 beta (IL-1 beta) in human fibroblasts and BC-8701 breast cancer cells. *J Cell Biochem,* 1997, 66, 322-36.

[6] Kirnbauer, R; Köck, A; Neuner, P; Forster, E; Krutmann, J; Urbanski, A; Schauer, E; Ansel, JC; Schwarz, T; Luger, TA. Regulation of epidermal cell interleukin-6 production by UV light and corticosteroids. *J Invest Dermatol,* 1991, 96, 484-9.

[7] Wlaschek, M; Heinen, G; Poswig, A; Schwarz, A; Krieg, T; Scharffetter-Kochanek, K. UVA-induced autocrine stimulation of fibroblast-derived collagenase/MMP-1 by interrelated loops of interleukin-1 and interleukin-6. *Photochem Photobiol,* 1994, 59, 550-6.

[8] Köck, A; Schwarz, T; Kirnbauer, R; Urbanski, A; Perry, P; Ansel, JC; Luger, TA. Human keratinocytes are a source for tumor necrosis factor alpha: evidence for synthesis and release upon stimulation with endotoxin or ultraviolet light. *J Exp Med.,* 1990, 172, 1609-14.

[9] Alexander, JP; Acott, TS. Involvement of protein kinase C in TNFα regulation of trabecular matrix metalloproteinases and TIMPs. *Invest Ophthalmol Vis Sci.,* 2001, 4, 2831-8.

[10] Johnson, GJ. The environment and the eye. *Eye,* 200, 18, 1235-50.

[11] Bloom, BR; Bennett, B. Mechanism of a reaction in vitro associated with delayed-type hypersensitivity. *Science,* 1966, 153, 80-2.

[12] David, JR. Delayed hypersensitivity in vitro: its mediation by cell-free substances formed by lymphoid cell-antigen interaction. *Proc Natl Acad Sci USA.* 1966, 56, 72-7.

[13] Weiser, WY; Temple, PA; Witek-Giannotti, JS., et al. Molecular cloning of a cDNA encoding a human macrophage migration inhibitory factor. *Proc Natl Acad Sci USA,* 1989, 86, 7522-6.

[14] Bernhagen, J; Calandra, T; Mitchell, RA; Martin, SB; Tracey, KJ; Voelter, W; Manogue, KR; Cerami, A; Bucala, R. MIF is a pituitary-derived cytokine that potentiates lethal endotoxaemia. *Nature,* 1993, 365, 756-9.

[15] Bucala, R. MIF re-evaluated: pituitary hormone and glucocorticoid-induced regulator of cytokine production. *FASEB J,* 1996, 7, 19-24.

[16] Calandra, T; Bernhagen, J; Mitchell, RA; Bucala R. The macrophage is an important and previously unrecognized source of macrophage migration inhibitory factor. *J Exp Med.,* 1994, 179, 1895-902.

[17] Lanahan, A; Williams, JB; Sanders, LK; Nathans, D. Growth factor induced delayed early response genes. *Mol Cell Biol.,* 1992, 12, 3919-29.

[18] Wistow, GJ; Shaughnessy, MP; Lee, DC; Hodin, J; Zelenka, PS. A macrophage migration inhibitory factor is expressed in the differentiating cells of the eye lens. *Proc Natl Acad Sci USA,* 1993, 90, 1272-5.

[19] Bacher, M; Meihnardt, A; Lan, HY; Mu, W; Metz, CN; Chesney, JA; Calandra, T; Bucala, R. Migration inhibitory factor expression in experimentally induced endotoxemia. *Am J Pathol,* 1997, 150, 235-46.

[20] Nishihira, J. Macrophage miration inhibitory factor (MIF): Its Essential role in the immune system and cell growth. *J Interferon Cytokine Res.,* 2000, 20, 751-62.

[21] Shimizu, T; Ohkawara, A; Nishihira, J; Sakamoto, W. Identification of macrophage migration inhibitory factor (MIF) in human skin and its immunohistochemical localization. *FEBS Lett.,* 1996, 381, 199-202.

[22] Watanabe, H; Shimizu, T; Nishihira, J; Abe, R; Nakayama, T; Taniguchi, M; Sabe, H; Ishibashi, T; Shimizu, H. Ultraviolet A-induced production of matrix metalloproteinase-1 is mediated by macrophage migration inhibitory factor (MIF) in human dermal fibroblasts. *J Biol Chem.,* 2004, 279, 1676-83.

[23] Shimizu, T; Abe, R; Nakamura, H; Ohkawara, A; Suzuki, M; Nishihira, J. High expression of macrophage migration inhibitory factor in human melanoma cells and its role in tumor cell growth and angiogenesis. *Biochem Biophy Res Commun,* 1999, 264, 751-8.

[24] Shimizu, T; Abe, R; Nishihira, J; Shibaki, A; Watanabe, H; Nakayama, T; Taniguchi, M; Ishibashi, T; Shimizu, H. Impaired contact hypersensitivity in macrophage migration inhibitory factor (MIF)-deficient mice. *Eur J Immunol,* 2003, 33, 1478-87.

[25] Shimizu, T. Role of macrophage migration inhibitory factor (MIF) in the skin. *J Dermatol Sci.,* 2005, 37, 65-73.

[26] Matsuda, A; Tagawa, Y; Matsuda, H; Nishihira, J. Identification and immunohistochemical localization of macrophage migration inhibitory factor in human cornea. *FEBS Lett.,* 1996, 385, 225-8.

[27] Matsuda, A; Tagawa, Y; Yoshida, K; Matsuda, H; Nishihira, J. Expression of macrophage migration inhibitory factor in rat retina and its immunohistochemical localization. *J Neuroimmunol,* 1997, 77, 85-90.

[28] Matsuda, A; Tagawa, Y; Matsuda, H; Nishihira, J. Expression of macrophage migration inhibitory factor in corneal wound healing in rats. *Invest Ophthalmol Vis Sci.,* 1997, 38, 1555-62.

[29] Wistow, GJ; Shaughnessy, MP; Lee, DC; Hodin, J; Zelenka, PS. A macrophage migration inhibitory factor is expressed in the differentiating cells of the eye lens. *Proc Natl Acad Sci USA.,* 1993, 90, 1272-5.

[30] Shimizu, T; Abe, R; Ohkawara, A; Nishihira, J. Ultraviolet B radiation up-regulates the production of macrophage migration inhibitory factor (MIF) in human epidermal keratinocytes. *J Invest Dermatol,* 1999, 112, 210-5.

[31] Ohnishi, Y; Tajima, S; Akiyama, M; Ishibashi, A; Kobayashi, R; Horii, I. Expression of elastin-related proteins and matrix metalloproteinases in actinic elastosis of s un-damaged skin. *Arch Dermatol Res.,* 2000, 292, 27-31.

[32] Fisher, GJ; Wang, ZQ; Datta, SC; Varani, J; Kang, S; Voorhees, JJ. Pathophysiology of premature skin aging induced by ultraviolet light. *N Engl J Med.,* 1997, 337, 1419-28.

[33] Brenneisen, P; Oh, J; Wlaschek, M; Wenk, J; Briviba, K; Hommel, C; Herrmann, G; Sies, H; Scharffetter-Kochanek, K. Ultraviolet B wavelength dependence for the regulation of two major matrix-metalloproteinases and their inhibitor TIMP-1 in human dermal fibroblasts. *Photochem Photobiol.,* 1996, 64, 877-85.

[34] Brauchle, M; Gluck, D; Di, Padova, F; Han, J; Gram, H. Independent role of p38 and ERK1/2 mitogen-activated kinases in the upregulation of matrix metalloproteinase-1. *Exp Cell Res.,* 2000, 258, 135-44.

[35] Wlaschek, M; Wenk, J; Brenneisen, P; Briviba, K; Schwarz, A; Sies, H; Scharffetter-Kochanek, K. Singlet oxygen is an early intermediate in cytokine-dependent ultraviolet-A induction of interstitial collagenase in human dermal fibroblasts in vitro. *FEBS Lett.,* 199, 413, 239-242.

[36] Honda, A; Abe, R; Makino, T; Norisugi, O; Fujita, Y; Watanabe, H; Nishihira, J; Yamagishi, S; Iwakura, Y; Shimizu, H; Shimizu, T. Interleukin-1β and macrophage migration inhibitory factor (MIF) in dermal fibroblasts mediate UVA induced matrix metalloproteinase-1 expression. *J Dermatol Sci.,* 2008, 49, 63-72.

[37] Hernandez-Pigeon, H; Jean, C; Charruyer, A; Haure, MJ; Baudouin, C; Charveron, M; Quillet-Mary, A; Laurent, G. UVA induces granzyme B in human keratinocytes through MIF: implication in extracellular matrix remodeling. *J Biol Chem.,* 2007, 282, 8157-64.

[38] Charman, WN. Ocular hazards arising from depletion of the natural atmospheric ozone layer: a review. *Ophthalmic Physiol Opt.,* 1990, 10, 333-341.

[39] Cejková, J; Stípek, S; Crkovská, J; Ardan, T; Midelfart, A. Reactive oxygen species (ROS)-generating oxidases in the normal rabbit cornea and their involvement in the corneal damage evoked by UVB rays. *Histol Histopathol,* 2001, 16, 523-33.

[40] Merriam, JC; Lofgren, S; Michael, R; Soderberg, P; Dillon, J; Zheng, L; Ayala, M. An action spectrum for UV-B radiation and the rat lens. *Invest Ophthalmol Vis Sci.,* 2000, 41, 2642-7.

[41] Reddy, VN; Giblin, FJ; Lin, LR; Chakrapani, B. The effect of aqueous humor ascorbate on ultraviolet-B-induced DNA damage in lens epithelium. *Invest Ophthalmol Vis Sci.,* 1998, 39, 344-50.

[42] Schein, OD. Phototoxicity and the cornea, *J Natl Med Assoc.,* 1992, 84, 579-583

[43] Ringvold, A; Davanger, M. Changes in the rabbit corneal stroma caused by UV-radiation. *Acta Ophthalmol (Copenh),* 1985, 63, 601-6.

[44] Pitts, DG; Bergmanson, JP; Chu, LW. Ultrastructural analysis of corneal exposure to UV radiation. *Acta Ophthalmol (Copenh),* 1987, 65, 263-73.

[45] Doughty, MJ; Cullen, AP. Long-term effects of a single dose of ultraviolet-B on albino rabbit cornea--II. Deturgescence and fluid pump assessed in vitro. *Photochem Photobiol,* 1990, 51, 439-49.

[46] Podskochy, A; Gan, L; Fagerholm, P. Apoptosis in UV-exposed rabbit corneas. *Cornea,* 2000, 1999-103.

[47] Lu, L; Wang, L; Shell, B. UV-induced signaling pathways associated with corneal epithelial cell apoptosis. *Invest Ophthalmol Vis Sci.,* 2003, 44, 5102-9.

[48] Wang, L; Dai, W; Lu, L. Ultraviolet irradiation-induced K(+) channel activity involving p53 activation in corneal epithelial cells. *Oncogene,* 2005, 24, 3020-7.

[49] Kozak, I; Klisenbauer, D; Juhas, T. UV-B induced production of MMP-2 and MMP-9 in human corneal cells. *Physiol Res.,* 2003, 52, 229-34.

[50] Kitaichi, N; Shimizu, T; Yoshida, K; Honda, A; Yoshihisa, Y; Kase, S; Ohgami, K; Norisugi, O; Makino, T; Nishihira, J; Yamagishi, S; Ohno, S. Macrophage migration inhibitory factor ameliorates UV-induced photokeratitis in mice. *Exp Eye Res.,* 2008, 86, 929-36.

In: Environmental Regulation: Evaluation, Compliance... ISBN: 978-1-60741-645-6
Ed: Diederik Meijer and Fillipus De Jong

Chapter 7

SPRING PROTECTION AGAINST MICROBIAL CONTAMINATION IN COMPARTMENTALIZED CARBONATE AQUIFERS, CENTRAL-SOUTHERN ITALY

Gino Naclerio*[*] and Fulvio Celico

Groundwater Research Center, Università degli Studi del Molise, Contrada Fonte Lappone, 86090 Pesche (IS), Italy.

ABSTRACT

Recent researches in central-southern Italy demonstrated that carbonate aquifers may behave as compartmentalized systems, due to the barrier action of some low-permeability fault zones that partially impede groundwater flow. This conceptual model is different from models generally applied in karst aquifers where faults act as drains, due to their enhanced hydraulic conductivity.

Due to this discontinuous heterogeneity, these aquifers behave as basin-in-series systems where seasonal springs occur along some fault zones. Thus, the groundwater generally flows towards a main perennial spring and a number of seasonal springs. Moreover, in some basin-in-series systems relationships between adjacent compartments may change during each hydrologic year, due to differences in head fluctuation over time. Hence, the basin that feeds a spring may behave as a non-static system, and its boundaries in high flow may be different from boundaries in low flow.

To date, no specific strategies have been defined by Italian Law to protect perennial and seasonal springs against pollution in such carbonate environments. The present chapter will focus the attention on the correct methodology to update and refine existing protection measures, with emphasis on microbiological contamination.

---

[*] Corresponding author: E-mail: naclerio@unimol.it, celico@unimol.it; tel: 0039 0874 404112; fax: 0039 0874 404123

INTRODUCTION

Carbonate aquifers provide the main drinking-water resources of central-southern Italy, supplying an average volume of about 10400 · 10^6 m^3 y^{-1} (Celico et al., 2000). Several important towns (i.e. Rome and Naples), many smaller towns and countless villages utilize this water for their public water supply. High altitude springs are often utilized by small towns and villages as well as by bottling plants. These springs are often affected by microbial pollution (Celico et al., 2004a; Celico et al., 2004b; Allocca et al., 2008; Naclerio et al., 2008) and protecting measures must be applied to prevent or minimize this contamination. However, due to the strategic importance of grazing and agriculture in carbonate massifs of central-southern Italy, it is unrealistic the application of a total protection. According to Biondic et al. (1998), scientists have to work to define a protection under sustained development conditions, taking into account the socio-economic needs of local inhabitants. This approach requires a high level of knowledge of carbonate aquifers hydrogeology and a thorough analysis of the interactions between land use and groundwater contamination.

The purpose of the present chapter is to refine actual protecting criteria of groundwater in carbonate aquifers of central-southern Italy, under sustained development conditions. These criteria have been refined taking into account: (a) the Italian Law and respective Addenda, which do not provide sufficient or effective solutions for problems in these hydrogeological settings, (b) the solutions recently proposed for the same environments by Celico et al. (2008), (c) the recent findings concerning the hydrogeological behaviour of such aquifer systems, and (d) the recent knowledge concerning the mechanisms that govern the spatial and the temporal evolution of microbial contamination of spring- and groundwater. Thus, we synthesize the main features of the studied hydrogeologic system and discuss about the main factors that influence the transport of microbial pollutants in the same environment, before focusing the attention on spring protection strategies.

HYDROGEOLOGICAL FRAMEWORK

Recently, the analysis of physical properties of fault rocks suggested a barrier action of some faults in platform carbonates in central Italy (Agosta et al., 2007; Billi et al., 2008), in agreement with the actual conceptual model (Celico, 1986) which was defined through hydrogeological investigations at regional or catchment scale (fault zones within the carbonate Italian Apennines may act as low-flow or no-flow boundaries). Celico et al. (2006) experimentally analyzed the hydrogeological behaviour of some fault zones in a test site in southern Italy, and demonstrated that they act as low-flow barriers. This conceptual model is different from models generally applied in karst aquifers where faults act as drains, due to their enhanced hydraulic conductivity. Thus, the studied faulted carbonate aquifers seem to behave as other aquifer types, such as the Triassic Sherwood Sandstone, where a significant compartmentalization was observed due to low-flow fault zones (Mohamed and Worden, 2006).

Despite the lower permeability, some fault zones allow significant groundwater flowthrough, and interdependence of hydraulic heads upgradient and downgradient of these faults has been observed during recession (Celico et al., 2006). Due to this discontinuous heterogeneity (sensu Freeze and Cherry, 1979), the aquifer behaves as a basin-in-series system where seasonal springs occur along some fault zones (Celico et al., 2006). Recent researches demonstrated that a basin-in-series system may behave as a non-static system where relationships between compartments may change over time. Relationships between compartments may change cyclically during each hydrologic year or episodically when dryer or wetter periods cause the hydraulic heads to reach anomalous equlibria.

The carbonate media are laterally and vertically well connected in the subsurface and the fracture spacing is often sufficiently dense to apply the continuum approach to describe groundwater flow also at the metric scale (Petrella et al., 2007). Darcy's law can be applied in epikarstic horizons with not pervasive karstification, and groundwater flow is expected to be laminar also in the underlying fractured bedrock (Petrella et al., 2008). Rainwater percolation is diffuse, mainly due to high fracture density and good interconnection of openings within limestone (Petrella et al., 2007). Since no retention of percolation and water storage in temporary perched aquifers are observed, hydraulic head and spring discharge fluctuations are directly influenced by distribution of precipitation over time (Petrella et al., 2007). This is different from findings in classic karst aquifers where temporary saturation can occur in the epikarst during heavy rainfall and hypodermic flow reaches the deeper saturated zone by rapid flow in open conduits (funnelling) and slow flow in a network of fine cracks (Drogue, 1992; Mangin, 1994).

MICROBIAL CONTAMINATION

Due to cattle grazing and manure spreading, carbonate aquifers in central-southern Italy are often affected by diffuse microbial contamination of groundwater (Celico et al., 2004a; Celico et al., 2004b; Allocca et al., 2008; Naclerio et al., 2008), and cases of gastroenteritis have often been reported and associated with such a pollution (Celico et al., 2004b). Recent studies (Celico et al., 2004b; Naclerio et al., 2008; Naclerio et al., 2009a; Naclerio et al., 2009b) demonstrated a great influence of pyroclastic soils on retention of fecal bacteria, therefore suggesting that they play an important role in protecting groundwater in such environments. In detail, pyroclastic soil influences (a) the distribution of fecal contamination over time, due to the diffuse interaction between microorganisms and soil medium at the first stage of microbial transport from the ground towards the groundwater, and (b) the amount of microbial cells that are transported towards the groundwater and then the springs, due to significant cells retention within the soil and the negligible role of carbonate rocks that are characterized by openings significantly wider than microbial cells dimension. Concerning retention, the organic matter of soil seems to play no role in retaining fecal bacteria (Naclerio et al., 2009a) and *Bacillus* spores (Naclerio et al., 2009b). Conversely, the clay fraction significantly influences retention of fecal bacteria (Naclerio et al., 2009a).

Such factors, together with the precipitation regime, govern the discontinuous distribution of microbial contamination over time in springs water. Such a distribution is emphasized in smaller aquifers, due to shorter pathways in both the unsaturated and the saturated media.

SPRING PROTECTION STRATEGIES

Taking into account the results of recent studies about microbial contamination of groundwater in compartmentalized aquifer systems within the carbonate Apennines, specific protecting criteria must be integrated to those introduced by Italian Law.

The integrative measures, partially identified by Celico et al. (2008), can be synthesized as follows: (a) introduction of new protection zones to integrate those indicated by Italian Law (zones Ia, IIa and IIIa), (b) identification of criteria for delimiting zones IIa, (c) introduction of the concept of "developing protection zones", (d) use of "dynamic" protecting measures, when polluting human activities are already existent within the protection zones, and (e) introduction of "expandable protection zones" where several springs coexist in a basin-in-series system.

Before describing the new protection zones, an overview on those indicated by the Italian Law. Zone Ia includes the area immediately adjacent to the tapping area, within a distance no less than 10 meters. This zone comprises the capture and the facilities needed for operation, service and guarding. No other activities are allowed. Zone IIa comprises the immediate hinterland of the tapping area (around the zone Ia), but the Law does not provide operative criteria in order to delimit this zone, nor does it offer any standard dimension. Zone IIIa comprises all the catchment area which keeps filling the groundwater source (in the case study it generally ranges from a few tens to several hundreds of km^2).

New protection zones have to be introduced (Celico et al., 2008) in the swallow hole areas (zones Ib, within a radius of 10 meters from the swallow holes) and within endorheic basins and catchment areas of streams which feed a carbonate aquifer (zones IIb and IIIb), in order to correctly evaluate the role of surface – groundwater interaction on transport of contaminants into the aquifer. Zones IIb (characterized by more restrictions) comprise those areas where the surface runoff infiltrates in the subsurface (a) into a swallow hole and / or (b) through a high permeability medium as well as areas where there is not a significant dilution of surface water into groundwater. Differently, zones IIIb (characterized by less restrictions) comprise those areas where the surface runoff infiltrates in the subsurface through a low permeability medium as well as areas where there is a significant dilution of surface water into groundwater.

Since no technical solutions are given by the Law, it was chosen the travel time criterion for delimiting zones IIa, taking into account the persistence of bacteria into aquatic environments. Nevertheless, due to the high flow and transport velocities within the studied aquifers, the choice of the travel time that must be used for delimiting the boundaries of zones IIa should be diversified as a function of water destination. That to avoid the delimitation of overextended protection zones, which negatively influence the socio-economic needs of an area. The lower the water treatment allowed, the wider the zones IIa have to become. Given an aquifer, the widest extension must be used for bottling mineral water, because no treatments are allowed.

When polluting human activities exist in the hinterland of the groundwater capture, the immediate effectiveness of protecting measures can be obtained by using "developing protection zones" (Celico et al., 2008). These zones must be integrated with technical measures which minimize the capture of polluted groundwater. Within these zones it is necessary to work towards a gradual introduction and extension of non-polluting or low-pollution technologies. These integrative technical measures will minimize the capture of polluted groundwater. They have been defined as "dynamic measures" because they correspond to groundwater management solutions and / or capturing criteria. This approach can be applied, for example, where springs are fed by both surface and groundwater. These sources are often contaminated by the rapid and episodic arrival of polluted surface water which infiltrates into swallow holes directly interconnected with springs. If no measures can rapidly remove the source of pollution, the problem can be solved by drilling an horizontal hole through the fracture network and, then, by capturing just the non contaminated groundwater. The effectiveness of this kind of capture is limited to those cases in which there is no significant percolation of surface water from karst conduits through the fracture pattern. When there is a significant and wide interaction between surface and groundwater the microbial pollution is detected at springs and into wells drilled within the fracture network.

In some settings, a basin-in-series system is characterized by several springs that flow at different altitudes. In such a scenario, the boundaries of protection zones may be different if considering just the perennial spring that represents the main drain of the system, or considering some or all other springs of the same system. To solve this problem, we can design "expandable protection zones" whose boundaries change with changing the number and the location of the springs we want to involve in the protection measures. This approach can also solve problems induced by changing relationships between compartments, with emphasis on those cases where the inversion of groundwater flowthrough along a low-permeability fault zone is not a cyclic phenomenon observable during each hydrologic year, but a phenomenon that depends on medium to long term changes in groundwater equilibria within an aquifer system.

Conclusion

The protection criteria synthesized in this chapter can be adapted, in part, to other kinds of aquifers, with emphasis on those characterized by a significant interaction between surface and groundwater.

Taking into consideration the vulnerability to pollution of an aquifer, in each protection zone it will be possible to define the land use prohibitions. This information can be synthesized in a "map of land use restrictions". This map (Celico et al., 2008) represents the graphic core of a GIS-based specific Decision Support System (sDSS) whose goal is, in part, the protection of water resources against pollution, taking into account the socio-economic needs of the territory. The sDSS is addressed to Public and Private Corporations who distributes and/or utilizes water resources, such as waterworks, bottling plants and thermal resorts. They typically require to consider a multitude of social, legal, economic, and hydrogeological factors, and the sDSS have great potential for improving the management of water resources.

REFERENCES

Allocca, V; Celico, F; Petrella, E; Marzullo, G; Naclerio, G. *Environ. Geol.,* 2008, 55, 277-283.

Biondic, B; Biondic, R; Dukaric, F. *Environ. Geol.,* 1998, 34, 309-319.

Celico, F; Celico, P; De Vita, P; Piscopo, V. *Hydrogéologie,* 2000, 4, 39-47.

Celico, F; Musilli, I; Naclerio, G. *Environ. Geol.,* 2004a, *46,* 233-236.

Celico, F; Varcamonti, M; Guida, M; Naclerio, G. *Appl. Environ. Microbiol.* ,2004b, 70, 2843-2847.

Celico, F; Naclerio, G. *Water Int.* 2005, *30*, 530-537.

Celico, F; Petrella, E; Celico, P. *Terra Nova*, 06, 18, 308-313.

Celico, F; Petrella, E; Naclerio, G. *Water Int.,*2007a, 32, 475-482.

Celico, F; Petrella, E; Naclerio, G. In *Groundwater Vulnerability Assessment and Mapping;* Witkowski, A. J; Kowalczyk A; Vrba J, Eds; Taylor & Francis: London, UK, 2007b, pp 177-190.

Celico, F; Petrella, E; Marzullo, G; Naclerio, G. *Water Int.,*2008, 33, *116-126.*

Naclerio, G; Petrella, E; Nerone, V; Allocca, V; De Vita, P; Celico, F. *Hydrogeol. J.,* 2008, 16, 1057-1064.

Naclerio, G; Nerone, V; Bucci, A; Allocca, V; Celico, F. *Colloid. Surf. B Biointerfaces,* 2009a, 72, *57-61.*

Naclerio, G; Fardella, G; Marzullo, G; Celico, F. *Colloid. Surf. B Biointerfaces,*2009b, 70, *25-28.*

Petrella, E; Capuano, P; Celico F. *Terra Nova.,* 007, 19, 82-88.

Petrella, E; Falasca, A; Celico, F. *Geofluids,*2008, 8, 159-166.

In: Environmental Regulation: Evaluation, Compliance... ISBN: 978-1-60741-645-6
Ed: Diederik Meijer and Fillipus De Jong

Chapter 8

ANTIGENIC/ALLERGENIC RUBBER PROTEINS AND ENVIRONMENTAL REGULATIONS

Michael J. Dochniak[1*] ***and Denise Hennes***[2]
[1]2653 RiceCreek Road #105, New Brighton, MN 55112, 1-952-808-941
[2]106 E. 125th Street, Burnsville, MN 55337, 1-952-808-0941

ABSTRACT

Health and safety issues associated with the antigenic/allergenic proteins inherent in natural latex continue to affect Government policy and practice. Although many of the issues associated with Hevea Brasiliensis natural rubber latex are well documented and understood, industrial society's magnificent exploitation and comfortable dependence on such a material continues to stress human health. Agencies including the Food and Drug Administration, the Environmental Protection Agency, and the Centers for Disease Control continue to develop and implement regulations, policies, and procedures for Hevea Brasiliensis natural rubber latex. Government agencies who have been emboldened to guide health and safety have shown contradictions in policy with regard to Hevea-Brasiliensis natural rubber latex. In the prevention of chronic disease and mental health disorders, we should question if such agencies should be in the business of the proliferation of natural rubber latex, without regard for the antigenic proteins therein. Finally, the increased incidence of allergic disease in industrialized societies presents the ultimate challenge for the environmental regulation of Hevea Brasiliensis natural rubber latex.

INTRODUCTION

Antigenic/Allergenic rubber proteins have likely stressed adaptive-immune systems throughout mankind's history. Exposure to rubber proteins by the exploitation of lacticiferous plants happened as early as 1600 B.C. Modern evidence has shown that the Mesoamericans, who had some of the most complex and advanced cultures including the Maya and the Aztec,

* Corresponding author: mdochniak@yahoo.com

used stabilized rubber from the plant Castilla elastica as early as 1600 B.C. In the early 19th century, natural rubber became an important raw material in industrialized societies when Charles Goodyear discovered a chemical process (vulcanization) that cured the rubber latex to remove its “stickiness”. Thereafter, natural rubber latex was transformed into a magnificent number of products. At present, it is estimated that more than 40,000 products contain natural rubber latex.

A major source of natural rubber latex comes from the Hevea Brasiliensis rubber tree. Unfortunately, inherent in Hevea Brasiliensis natural rubber latex are antigenic/allergenic proteins that often trigger adverse immune responses (i.e., allergies) in humans. Hevea Brasiliensis contains about 2-5 percent protein by weight. Analysis indicates that there are about 200 dissimilar proteins therein and about 50-60 of these proteins are suspected allergens. The World Health Organization – International Union of Immunological Societies has assigned names to about 13 of these allergens. Furthermore, published data indicate that up to 6 percent of the general public and 18 percent of health care workers have some level of natural rubber latex allergy. While studies repeatedly uncover high prevalence rates, the nonspecific nature of symptoms and lack of knowledge about natural-latex allergy result in missed diagnosis in many sensitized persons who are at risk of progression to serious allergic reactions. In 2008, Johns Hopkins Hospital has banned nearly all Hevea-Brasiliensis natural rubber latex products. [1]

It is well known that exposure to the Hevea-Brasiliensis natural rubber latex proteins can cause chronic disease. These proteins are considered chronic non-infectious agents or allergens, factors that affect immune susceptibility to such proteins include repeated exposure and genetic predisposition. Repeated exposure to these allergens through inhalation and/or dermal absorption has been shown to cause an increased incidence of sensitization, adverse allergic reactions, and death through anaphylactic shock. For example, health care workers have experienced an increased number of sensitization's and deaths from natural-rubber-based gloves in an attempt to reduce the accidental spread of viral infections including Autoimmune Deficiency Syndrome. Because of such worker exposure data, Hevea-Brasiliensis natural rubber latex is currently recognized as a hazardous material by the National Institute for Occupational Safety and Health. [2]

Children are often exposed to products formed from Hevea-Brasiliensis natural rubber latex including nipples for baby bottles, pacifiers, diaper adhesives, mattresses, bed pads, balloons, teething articles, toys, rubber bands, pencil erasers, carpet padding, food packaging adhesives and septum-capped medicinal vials used for vaccines. Consumer groups are calling for warning labels on food packaging containing latex, saying the substance poses a potential threat to people with allergic sensitivities [3] and John Hopkins Researchers have encouraged the Food and Drug Administration and Pharmaceutical companies to discontinue the use of Hevea-Brasiliensis natural rubber latex stoppers in medical vials. [4] A single exposure may cause an immune response and a study investigating natural rubber latex allergy in children concluded that the prevalence is about 1 percent among atopic children. [5] It has often been said that latex allergy usually affects people who are routinely exposed to natural rubber latex. Unfortunately, many neonates are routinely exposed to bottle nipples and pacifiers that are formed from Hevea Brasiliensis natural rubber latex. Currently, we are at the mercy of rubber manufacturers; many of which are located in third world countries.

An immune response associated with the Hevea-Brasiliensis natural rubber latex proteins can induce immunoglobulin-E secreting lymphocytes such as plasma B-cells and memory B-cells to form antibodies that target dissimilar exogenous/endogenous proteins through a cross-react mechanism. Compared to other antibodies free-floating in the blood stream, immunoglobulin-E has low concentration (~ 0.001%) and a short half-life (~ 2-days). Immunoglobulin-E antibodies are specific to a select group of antigens, rather than a wide range, and are extremely biologically active despite low concentrations in circulation. This is because Immunoglobulin-E antibodies bind to high-affinity receptors on the surface of mast cells and basophiles, so that these cells may be highly sensitive to antigens even when the concentration of immunoglobulin-E in the circulation is very low. Thus, immunoglobulin-E plays a major role in reactivity to the antigenic/allergenic protein in Hevea-Brasiliensis natural rubber latex.

Mammals are the only organisms that produce the antibody immunoglobulin-E associated with adaptive immunity. In humans, immunoglobulin-E plays a primary role in allergic responses resulting in a cascade of chemical reactions including neurotransmitter release and cross-react immune responses that affect the incidence of disease.

Chronic allergic disease and mental health disorders have been associated with Hevea-Brasiliensis natural rubber latex allergy. It is known that proteins in foodstuff that are homologous to the Hev-b proteins may cross-react with immunoglobulin-E antibodies that have been formed from a Hevea-Brasiliensis natural rubber latex allergy. One study showed that almost 50 percent of patients with Hevea-Brasiliensis natural rubber latex allergy showed food allergy. [6] Natural-latex allergy may be a catalyst for the onset of food allergies. What can be perplexing is Hevea-Brasiliensis natural rubber latex allergy can go into remission from reduced exposure, while subsequent food allergies there from may remain persistent based on repeated exposure to such proteins. Thus, the after effects of Hevea-Brasiliensis natural rubber latex allergy may continue to stress adaptive immunity long after the natural-latex allergy has gone into remission.

The degree of Hevea-Brasiliensis natural rubber latex allergy and immunoglobulin-E selectivity there from may play a critical role in the onset of mental health disorders. In human cognition studies, it has been speculated that natural-latex allergy may affect the incidence of Schizophrenia. [7] In parallel, one of the five pervasive developmental disorders officially recognized by the American Psychiatric Association's *Diagnostic and Statistical Manual of Mental Disorders* (DSM-IV) is childhood disintegrative disorder, initially termed childhood schizophrenia. Furthermore, it has been suggested that Hevea-Brasiliensis natural rubber latex exposure during prenatal/neonatal development may have affected the incidence of atopy and allergy induced autism. [8]. In a recent study entitled, *Allergic manifestations in autistic children: Relations to disease severity,* researchers concluded that allergy may play a role in the pathogenesis of autism wherein allergic immune responses to some proteins (e.g., dietary protein and natural latex) may induce the production of brain auto-antibodies, which are found in many autistic children. [9].

With such an understanding of the health and safety issues associated with natural-latex we can ask the question: Why does Hevea-Brasiliensis natural rubber latex continue to stress human health?

In the book *Silent Spring*, wildlife biologist Rachel Carson taught that chemicals that become universally common or repetitive can assume "the harmless aspect of the familiar". Even though many of the health and safety aspects of Hevea-Brasiliensis natural rubber latex are well understood, industrial society's magnificent exploitation and comfortable dependence on such a material continues to stress human health based on a inertia sustained the *the harmless aspect of the familiar.* More importantly, Government agencies that have been emboldened to guide health and safety have shown contradictions in policy for imported virgin feed-stock and recycled Hevea-Brasiliensis natural rubber latex.

In 2004, the United States Product Safety Commission denied petition HP00-2 requesting a rule declaring Hevea-Brasiliensis natural rubber latex to be a strong sensitizer. The Honorable Thomas H. Moore (Commissioner) stated, "Nevertheless, it would behoove manufacturers of NRL to take steps to reduce the level of proteins that consumers can come into contact with, whether or not the end product is a medical device. [10]

The United States Environmental Protection Agency (EPA) is involved in the promotion of recycled natural rubber latex. The use of crumb rubber from recycled tires has been touted by the EPA and a number of state environmental agencies as having beneficial impact on the environment. [11] Furthermore the EPA promotes the use of ground rubber tires, without regard for the antigenic/allergenic protein therein, in recreational applications and states, "The material can absorb much of the impact from falls providing added safety for children". [12]

In United States Patent 6,407,144 the Government is involved in the recycling of natural rubber latex without regard for the antigenic/allergenic proteins therein. In the Government Interests section of the patent it states, "The United States Government has rights to this invention pursuant to contract number DE-AC09-96-SR18500 between the United States Department of Energy and Westinghouse Savannah".

In 2008, a citizen petition was filed under section 21 of the Toxic Substance Control Act (TSCA) requesting that the Director issue a regulation that prohibits the use and distribution in commerce of Hevea-Brasiliensis natural rubber latex adhesives having total protein content greater than 200-micrograms per dry weight of latex. The petitions intent was that implementation of an EPA regulation that guides adhesive manufacturers to use Hevea-Brasiliensis natural rubber latex that satisfy such requirements may affect the incidence of latex allergy and allergy induced autism in neonates. The petition was denied and in the deposition the Assistant Administrator stated, "The petition does not present facts establishing that natural latex adhesives containing any specific level of protein present an unreasonable risk". [13] It is well known that there are established companies that have demonstrated that allergenic proteins can be substantially removed (e.g., < 20 micrograms total protein per gram dry latex) from Hevea Brasiliensis latex to provide adhesives having enhanced non-allergenicity and effective bonding characteristics. [14]

On April 23, 2008, the Food and Drug Administration cleared for marketing the Yulex Patient Examination Glove which is the first medical device made from guayule latex, a new form of natural rubber latex having reduced allergenic protein content.

CONCLUSION

In the prevention of chronic disease we should question if the non-regulation of virgin Hevea-Brasiliensis natural rubber latex and recycled natural latex there from, without regard to the antigenic/allergenic protein therein, is good Government policy when considering the health and safety of all citizens.

REFERENCES

[1] Physorg.com. Latex banned at Johns Hopkins Hospital. January 18, 2008. Available from: http://www.physorg.com/news119886779.html

[2] NIOSH Publication No. 97-135. Preventing Allergic Reactions to Natural Rubber Latex in the Workplace. June 1997 Available from: http://www.cdc.gov/niosh/latexalt.html

[3] Science Daily. Deadly Latex Evading. *Lax Food Labeling Laws*. August 9, 2006. Available from: http://www.sciencedaily.com/releases/2006/08/060809083433.htm

[4] Kate, O'Rourke. Drug Bottles Containing Natural Rubber Stoppers May Place Latex Allergic Patients at Risk for Reactions: Hopkins Researchers Encourage FDA and Pharmaceutical Companies to End Natural Rubber Stopper Use. June 8, 2001. Available from: http://www.hopkinsmedicine.org/press/2001/JUNE/010608.htm

[5] Ylitalo Leea. Natural rubber latex allergy in children. January 22, 2000. Available from: http://acta.uta.fi/teos.phtml?3312

[6] Carlos Blanco, Latex-Fruit Syndrome, *Current Allergy and Asthma Reports* 2003, 3, pp 47-53.

[7] Harold D. Foster, Schizophrenia: The Latex Allergy Hypothesis. *Journal of Orthomolecular Medicine*, 1999, Vol. 14, No.2, pp 83-90.

[8] Michael J. Dochniak, Autism Spectrum Disorders – Exogenous Protein Insult, *Medical Hypotheses*, 2007, 69(3), 545-549.

[9] Gehan A. Mostafa et al., Allergic manifestations in autistic children. *Relation to disease severity*, 2008, volume 6, Number 2, pp. 115-123.

[10] Todd Stevenson. Citizen Petition HP00-2. Requesting a Rule Declaring Natural Rubber Latex to be a Strong Sensitizer. April 30th, 2004. Available from: U.S. Consumer Product Safety Commission, deposition for HP00-2, 2004.

[11] Environmental Protection Agency. Waste-Resource Conservation –Common Wastes & Materials – Scrap Tires. Available from: http://www.epa.gov/osw/conserve/materials/tires/

[12] AstroTurf. General Sports Venue/AstroTurf USA's position on concerns over the use of crumb rubber derived from recycled passenger tires in synthetic turf systems. 2008 Available from: http://www.astroturfusa.com/news/Rubber.html

[13] Environmental Protection Agency. Natural Rubber Latex Adhesives; Disposition of TSCA Section 21 Petition. Federal Registrant: June 9th, 2008 (Volume 73, Number 111). Pages 32573-32577. Available from: http://edocket.access.gpo.gov/2008/E8-12850.htm

[14] Vytex – Natural Rubber Latex. Vytex Natural Rubber Latex (NRL) redefines the future of Latex. 2008. Available from: http://www.vytex.com/

In: Environmental Regulation: Evaluation, Compliance... ISBN: 978-1-60741-645-6
Ed: Diederik Meijer and Fillipus De Jong

Chapter 9

GENETIC AND ENVIRONMENTAL REGULATION AND ARTIFICIAL METABOLIC MANIPULATION OF ARTEMISININ BIOSYNTHESIS*

***Qing-Ping Zeng*[1*], *Rui-Yi Yang*[1,] *Li-Ling Feng*[2] *and Xue-Qin Yang*[2]**
[1]Laboratory of Biotechnology, Tropical Medicine Institute
[2]Artemisinin Research Center; Guangzhou University of Chinese Medicine
Guangzhou 510405 China

ABSTRACT

Artemisia annua is currently sole herbaceous biomass for industrial manufacture of artemisinin, an antimalarial sesquiterpene lactone with the unique endoperoxide architecture. Due to presence in trace amount, artemisinin has been targeted for *in planta* overproduction by genetic modification of *A. annua*. Beneficial from such pursuits, the overall enzymatic cascades involving artemisinin biogenesis have been elucidated and a dozen of critical artemisinin responsible genes identified. Consequently, transgenic *A. annua* plants with enhanced artemisinin production are available although substantial and profound potentials in artemisinin accumulation expected. Alternatively, due to conservation of the entire terpene pathways among higher plants and eukaryotic or even prokaryotic microbes, re-establishment of extended or diverted pathways toward *de novo* microbial artemisinin production has been eagerly attempted in genetically tractable microbes. In such aspect, a suit of downstream pathway genes specific for artemisinin biogenesis have been transplanted from *A. annua* into *Sacchromyce cerevisiae* and *Escherichia coli*, in which a series of incredible amounts of artemisinin precursors manufactured. The next-step goal is to further accelerate forward the total artemisinin biosynthesis through biotransformation of the artemisinin precursor(s) either *in vivo* or *in vitro*. For this purpose, the putative oxidant sink molecule capable of quenching the

* A version of this chapter was also published in *Biotechnology: Research, Technology and Applications*, edited by Felix W. Richter published by Nova Science Publishers, Inc. It was submitted for appropriate modifications in an effort to encourage wider dissemination of research.

* Corresponding author: E-mail: qpzeng@gzhtcm.edu.cn

reactive oxygen species (ROS), in particular, the singlet oxygen ($^1O_2$), must be produced, in a large scale, in genetically modified microbes or transgenic *A. annua* plants. Whether dihydroartemisinic acid or artemisinic acid is such a $^1O_2$-scavenging direct intermediate has not been convinced, but conversion from dihydroartemisinic acid or artemisinic acid to artemisinin recognized as a bottleneck for artemisinin biosynthesis and versatile strategies aiming at breaking the rate-limited step enthusiastically pursued in *A. annua*, for example, by utilization of the primary abiotic or biotic stress signals or secondary stress signal transducers. These achievements should benefit our future intervention with the homeostatic tempo-spatial regulation mode of genetic background-based and environment-dependent artemisinin accumulations. This article introduces, from the genomics, transcriptomics, proteomics and metabolomics, the updated literatures describing the relationship between artemisinin biosynthetic gene overexpression and subsequent artemisinin overproduction as well. It should shed light on further elucidation of the intrinsic rule and mechanism underlying that artemisinin biochemical synthesis is fine-tuned by the genetic and environmental regulators, and should also urge the researchers all over the world more intensively investigating the intriguing *A. annua* plant that has implications in the medicinal and aromatic industries.

Keywords: *Artemisinin annua*, Artemisinin, Environmental induction, Gene expression, Microbial engineering, Signal transduction, Transgenic plant.

1. INTRODUCTION

Artemisia annua, the Latin nomenclature of Qinghao in Chinese and annual wormwood or sweet wormwood in English, is a well-known medicinal herbage for chemically extracting antimalarial artemisinin, from which a series of therapeutically efficient artemisinin derivatives, including artemether, arteether, artesunate and artelinate were artificially semi-synthesized (Figure 1).

Nowadays, these improved artemisinin analogs have been recommended by the World Health Organization (WHO) as the essential component for artemisinin-based combination therapies (ACTs) in order to combat the multi-drug resistant malaria occurring in malarial endemic districts (WHO, 2001). For such consideration, three kinds of artemisinin derivetives, artesunate and artemether have been described in the *International pharmacopoeia* (WHO, 2003) and listed in the *WHO Model List* of *Essential Medicines* (WHO, 2005).

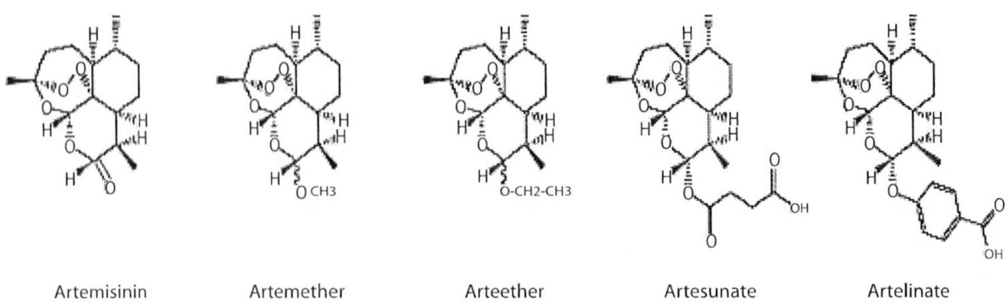

Figure 1. The structure of naturally isolated artemisinin and artificially semi-synthesized artemisinin derivatives.

Based on the thousand-year experiences and practices in ancient China for treatment of fever probably caused by malaria, Chinese scientists pioneered extensive search for antimalarial compound(s) in *A. annua* and first purified the antimalarial monomer, artemisinin, a molecule known as qinghaosu or arteannuin as an original nomenclature, at that time, and subsequently identified artemisinin as an intriguing type of the sesquiterpene lactone with a unique endoperoxide bridge (Liu et al. 1979). During that period, artemisinin was analyzed by the nuclear magnetic resonance (NMR), mass spectroscopy (MS) and X-ray crystal diffraction, from which the molecular weight (282.3), chemical formula ($C_{15}H_{22}O_5$) and configuration of artemisinin were successfully revealed.

As a natural secondary metabolite, artemisinin seems to serve as a phytoalexin that confers resistance to the plant itself against pathogenic microorganisms (Stoessl et al. 1976) and attraction to protective insects (Kappers et al. 2005) as other volatile terpenes do. Moreover, it has been recently suggested that artemisinin is most likely as a scavenger of the reactive oxygen species (ROS) to protect cells from damage caused by oxidative stresses (Wallaart et al. 2001).

In broad definition, the term "stress" indicates all extreme environmental conditions that induce generation of ROS. The natural stress circumstances are classified as the biotic stresses (disease and pest ingression and mechanical wounding, etc.) and abiotic stresses (cold, heat, irradiation, anoxia, drought, saline and alkaline, etc (Xiong et al. 2002). Although oxidative stress-induced overproduction of plant secondary metabolites has been documented in *Oryza sativa* (Nojiri et al. 1996; Tamogami et al. 1997), *Catharanthus roseus* (Menke et al. 1999), *Arabidopsis thaliana* (Brader et al. 2001), and *Cupressus lusitanica* (Zhao and Sakai, 2001), investigation regarding the relationship between oxidative stress and artemisinin biosynthesis are just thriving, and available data from these researches are fragmentary and incomplete. Nevertheless, the entire physiochemical process involving generation, reception and transduction of the oxidative stress signals, activation of the transcription factors and their binding to promoters, and induced expression of target genes has been initiated and may soon lead to validation for the hypothesis of oxidative stress induction as a rate-limited step in artemisinin biosynthesis (Zeng et al. 2008a).

This article prospectively summarize the cutting-edge technical frontier and the updated academic viewpoints that intend to attract and encourage global researchers dedicating their endeavors to exploration on the molecular mechanism of artemisinin and other valuable terpenes, and to stimulate their interests and enthusiasms in the understanding of oxidative stress-induced up-regulation of the artemisinin responsible genes as well as other plant secondary metabolism implicated in agriculture, medicine and industry.

2. An Overview on Species Character and Genetic Background of *A. annua*

2.1. Genome Organization and Phylogeny of *Artemisia* Genus

Torrell and Vallès (2001) estimated the genome size of *Artemisia* genera among five subgenera that compose 21 species and three subspecies and summarized their life forms, DNA amounts, ploidy states and karyotypes (Table 1).

Table 1. DNA content and karyological characters of *Artemisia* populations

Taxon	Life cycle	DNA content (pg)	Ploidy level (2n)	Chromosome length (µm)
Subgenus *Seriphidham*		8.80		
A. fragrans	P	5.35	18, 2x	24.84
A. caerulescens	P	6.66	18, 2x	30.17
A. herba-alba subsp. *valentina*	P	6.57	18, 2x	36.16
A. herba-alba subsp. *herba-alba*	P	12.48	36, 4x	56.45
A. barrelieri	P	12.96	36, 4x	65.36
Subgenus *Artemisia*		7.15		
A. annua	A	3.50	18, 2x	19.58
A. tournefortiana	A/B	6.69	18, 2x	
A. vulgaris	P	6.08	16, 2x	29.51
A. vulgaris	P	9.74	34, 4x	44.27
A. chamaemelofolia	P	6.04	18, 2x	27.77
A. molinieri	P	5.96	18, 2x	26.45
A. lucentica	P	7.68	16, 2x	27.33
A. Judaica	P	11.52	16, 2x	43.44
Subgenus *Absinthium*		11.26		
A. absinthium	P	8.52	18, 2x	38.46
A. thuscula	P	10.52	18, 2x	38.33
A. numelliformis	P	12.41	34, 4x	63.38
A. splendens	P	13.59	32, 4x	58.64
Subgenus *Dracunculus*		13.67		
A. campestris	P	5.87	18, 2x	25.29
A. campestris	P	11.00	36, 4x	47.18
A. monosperma	P	11.02	36, 4x	64.61
A. crithmifolia	P	15.60	54, 6x	81.92
A.dracunculus	P	23.22	90, 10x	-
Subgenus *Tridentae*		16.91		
A. tridentatae subsp. *spiciformis*	P	8.18	18, 2x	-
A. cana	P	25.65	72, 8x	-

Note: A: annual; B: biennial; P: perennial.

Although multiple ploidy levels of 2x, 4x, 6x, 8x and 10x are presented in *Artemisia*, two basic chromosome numbers, x = 9 and x = 8, were proposed for the genus, in which the latter chromosome sets may be derived from the former chromosome sets by a possible descendent chromosome fusion-based dysploidy mechanism (Vallès and Siljak-Yakovlev 1997). Later, Kreitschitz and Vallès (2003) counted the chromosome numbers of five *Artemisia* species pooled from Poland and identified the diploid *A. annua* (2n=18), tetraploid *A. absinthium* and *A. abrotanum* (2n=36), and decaploid *A. dracunculus* (2n=90) in Polish *Artemisia* species.

The first report on the cytological description of *A. annua* chromosomes appeared in 1928, when Weinedel-Libebau recognized the basic chromosome numbers (x=9) and somatic diploid (2n=18) among this species. Subsequently, more investigators provided evidence supporting that *A. annua* plants possess diploid chromosomes although the plants under investigation distribute in distinct geological areas in the World, including Hungary (Polya 1949), Japan (Suzuka 1950; Arano 1964), Bulgaria (Kuzamanov et al. 1986), Spain (Valles 1987), the south part of far east region in former Soviet Union (Vokova and Boyco 1986) and the inner Mongolia district of northern China (Fu 1991).

Wang et al. (1999) counted the chromosomes and analyzed the karyotypes of five species growing in the northeast provinces of China and belonging to *Abrotanum* sector of *Artemisia* genus, *A. tanacetifolia*, *A. annua*, *A. adamsii*, *A. palustris* and *A. aurata*. Their investigation indicated that these species are diploid with the common chromosome numbers (2n=18). The somatic chromosome morphology and karyotype of these five *Artemisia* species are illustrated in Figure 2.

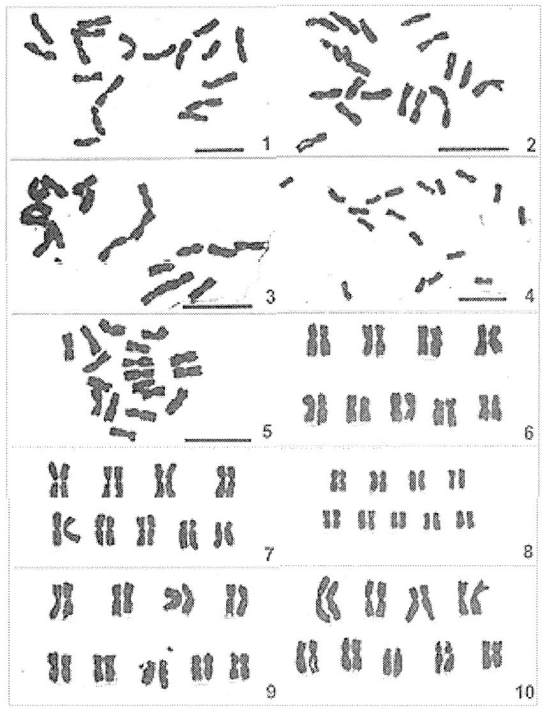

Figure 2. The morphology of somatic chromosome and karyotype in the five species of Artemisia. 1 and 7. A. tanacetifolia; 3 and 9. A. annua; 2 and 10. A. adamsii; 4 and 8. A. palustris; 5 and 6. A. aurata. Bar=10 μm. Adopted from Fu et al. (1991).

The tribe Anthemideae (Asteraceae) was classified into one subtribe Artemisiinae and other outgroup genera. The subtribe Artemisiinae includes the *Artemisia* subgenera (*Artemisia*, *Absinthium*, *Dracunculus*) and *Seriphidium* (new world *Seriphidium* and old world *Seriphidium*), which altogether comprise the *Artemisia* group. Watson et al. (2002) described the molecular phylogeny of subtribe Artemisiinae by comparison of the internal transcribed spacers (ITS) of nuclear ribosomal DNA (rDNA). The resulting evolutionary tree is comprised of three major clades that correspond to one radiate genus and two clades of *Artemisia* species. The sequences of ITS1 and ITS2 of rDNA were also analyzed by Vallès et al. (2003). Among 44 *Artemisia* species that represent five classical subgenera, the established eight main clades definitely supported the monophyly of *Artemisia* genus.

2.2. Morphological and Ecophysiological Traits of *A. annua*

A. annua, belonging to the Asteraceae (Compositae) family, is a species of aromatic annual herbal plant that originally distributes in the high altitude steppe region of north China (40_ N, 109_ E and 1000 to 1500 m above the sea level) (Wang 1961). However, the herbal plant now cosmopolitanly grows wild in the World and thrives in Asian, European and American countries (Lin and Lin 1991). Although the tropical Africa seems not suitable to cultivate the plant, large-scale plantation of *A. annua* has been in trials in Kenya and Tanzania.

As a short-day plant with a photoperiod between 12 and 16 h (13.5 h in average and 1000 h in total annual light hours), *A. annua* is very sensitive to short-day light stimulation, by which the branched plants will bloom after two weeks of such light induction. *A. annua* is reproduced by seeds that germinate on the temperature above 7°C and grow at 13.5-17.5°C with the total annual temperature needs of 3500-5000°C (Paniego et al. 1993). The whole plant of *A. annua* compose a single stem (2-6 mm in diameter) with alternative branches reaching 30-100 cm (more than 200 cm for cultivars) in height and deeply dissected leaves ranging from 3–5 cm in length and 2-4 cm in width. The aerial surface of leaves and flowers are covered with multi-celled biseriate glandular trichomes that accumulate mono- and sesquiterpenes (Duke et al. 1987). The glandular trichomes are comprised of ten cells that differentiate into five cell pairs with each cell pair apparently having different functions as deduced from the ultrastructural difference (Duke and Paul 1993). By the light microscopy, the basal, stalk and apical cell pairs are colorless, whereas the two subapical cell pairs are green (Figure 3).

Duke et al. (1994) concluded that artemisinin is sequestered in glandular trichomes of *A. annua* because flowers and leaves that store artemisinin and essential oils own abundant glandular trichomes, while those in a special *A. annua* biotype without or with scarce glands contains no or low artemisinin. Artemisinin is present in *A. annua* in a small amount range from 0.01 to 0.8 % of dry weight (Abdin et al. 2003). The time-course determination of artemisinin content at the different developmental stages revealed a positive correlation between plant growth stages and artemisinin yields, which was attributed to gradual expansion of the leaf surface and progressive increase of the biomass weight during plant growth (Singh et al. 1988). Furthermore, artemisinin content of inflorescence in the bud stage is not higher than in leaves, but it is four- to 11-fold higher in flowers at full bloom than in leaves (Ferreira et al. 1995). Some researchers indicated that artemisinin content is highest just before flowering (Acton and Klayman 1985; Liersh et al. 1986; ElSohly et al. 1990), but others observed that a peak artemisinin yield is reached during full flowering (Pras et al. 1991; Morales et al. 1993; Ferreira et al. 1995).

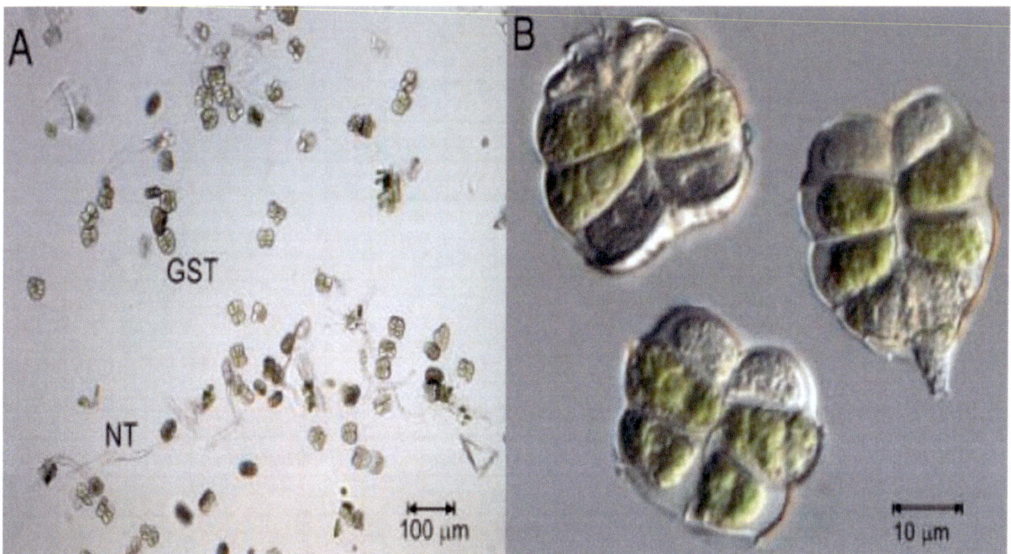

Figure 3. The microscopic morphology of glandular trichomes isolated from the floral bud of *A. annua*. (A) low magnification micrographic view; (B) high magnification phase contrast micrographic view. GST, glandular secretory trichome; NT, nonglandular trichome. Adopted from Covello et al. (2007).

The most abundant type of sesquiterpenes in *A. annua* is artemisinic acid, which occurs in an eight- to tenfold higher concentration (Roth and Acton 1989; Jung et al. 1990) and followed by arteannuin B (Klayman 1993). The fluctuation of artemisinin and the related sesquiterpene intermediates, such as artemisinic acid, arteannuin B, and artemisitene, in *A. annua* during the vegetation period was studied in Vietnamese varieties (Woerdenbag et al. 1994). The highest levels of artemisinin (0.86% dry weight), artemisinic acid (0.16% dry weight) and arteannuin B (0.08% dry weight) were measured in the leaves of five- month-old plants that possess highest leaf weight.

Recently, Lommen et al. (2007) found that the concentration levels of total artemisinin plus its precursors that include dihydroartemisinic acid, dihydroartemisinic aldehyde, dihydroartemisinic alcohol, artemisinic acid, artemisinic aldehyde, and artemisinic alcohol are higher in green leaves than in dead leaves in the younger stage, but are comparable at the final harvest stage. This means that conversion of the precursors to artemisinin is more advanced in dead leaves than in green leaves, which leads to the concentration of artemisinin being higher in dead leaves than in green leaves.

2.3. Chemical Constituents of *A. annua* Implicated in Medical and Fragrant Industries

The extensive searches for active compounds in *A. annua* have led to isolation of more than 150 kinds of secondary metabolites, including terpenoids, flavonoids, coumarins, steroids, phenolics, purines, lipids and aliphatic compounds (Bhakuni et al. 2001). The reported olefinic monoterpenes, sesquiterpenes, diterpenes and triterpenes are alphabetically listed in Table 2.

Table 2. The identified terpenoids and isolated terpene intermediates in *A.annua*

Terpenoids	Terpenoids
abscisic acid	chrysanthenone
abscisic acid methyl ester	cineol, 1-4
amorpha-4, 11-diene	cineol, 1-8
amyrenone, α	**copaene, α**
amyrin, α	copaene, β
amyrin, β	cymene, para
amyrin acetate, β	decan-2-one
annuic acid, *nor*	elemene, β
annulide	eudesma-4(15)-11-diene, 5-α-hydroxy
annulide, iso	**farnesene, β**
artemisia alcohol	**farnesene, β trans**
artemisia ketone	farnesene, *trans*-β
artemisia ketone, iso	fenchol
artemisinic acid	fenchone
artemisinic acid, dihydro	friedelan-3-beta-ol
artemisinic acid hydroperoxide, dihydro	friedelin
artemisinic alcohol	germacrene A
artemisinic alcohol, dihydro	germacrene D
artemisinic aldehyde	germacrene, bicyclo
artemisinic aldehyde, dihydro	**guaiene, α**
arteannuin A(Qinghaosu I)	hentriacontan-1-ol-triacontanoate
arteannuin B (Qinghaosu II)	hepta-3-trans-5-diene-2-one, 6-methyl
arteannuin B, deoxy (Qinghaosu III)	hept-2-ene, bicyclo (3, 1, 1), 3-7-7-trimethyl
arteannuin B, dihydro	hex-2-en-al
arteannuin B, dihydrodeoxy	hexcis-3-en-1-ol
arteannuin B, dihydro-*epi*-deoxy	hexacosan-1-ol
arteannuin B, *epi*-deoxy	hexadecanoic acid ethyl ether
arteannuin C	hexan-1-ol acetate
arteannuin D (Qinghaosu IV)	hexan-1-ol, 2-ethyl
arteannuin E (Qinghaosu V)	humulene, α
arteannuin F	limonene
arteannuin G	linalool
arteannuin H	linalool acetate
arteannuin I	longipinene
arteannuin J	menthen-4-ol, para
arteannuin K	menthol
arteannuin L	menthol, 2-hydroxy
arteannuin M	muurola-4,11-diene

Table 2. (Continued)

Terpenoids	Terpenoids
arteannuin N	myrcene
arteannuin O	myrcene, α-hydroperoxide
artemisin	myrcene, β-hydroperoxide
artemisinic acid methyl ester	myrtenal
artemisinic acid, α-epoxy	myrtenol
artemisinic acid, dihydro	nerolidol
artemisinic acid, 6, 7-dehydro	nortaylorione
artemisinic acid, dihydro, methyl ester	octan-1-ol
artemisinin (qinghaosu)	oleanolic acid
artemisinin, dehydro	phytene-1-2-diol
artemisinin, deoxy	phytol, *trans*
artemisinol	pinene
artemisitene	pinene, α
bisabolene, β	pinene, β
borneol	pinocamphone
borneol, acetate	pinocarveol, trans
cadina-4(15)-11-dien-9-one	sabiene
cadina-4(7)-11-dien-12-al	sabiene, cis-hydrate
cadinane, *seco*	selina-4,11-diene
cadinane, dihydroxy	selinene, β
cadinene, γ	taraxasterone
cadin-4-en-11-ol, 3-iso-butyryl	taraxerol acetate
cadin-4-ene, 3-α-7-α-dehydroxy	terpinen-4-ol
camphene	terpinene, α
camphene hydrate	terpinene, γ
camphor	terpineol, α
cineole, 1, 8	thujene, α
caryophyllene, α	thujone
caryophyllene, β	thujone, α
caryophyllene, oxide	thujone, iso
caryophyllene, trans	Trycyclene
cedrol	Ylangene

The content of essential oil in *A. annua* is generally as 0.02-0.49% (on a fresh weight basis) or 0.04-1.9% (on a dry weight basis), but depends on its geographical origin. Woerdenbag et al. (1993) detected 4.0% and 1.4% essential oil (V/W) from Chinese and Vietnamese varieties, respectively. The Chinese varieties-originated essential oil includes *artemisia* ketone (63.9%), *artemisia* alcohol (7.5%), myrcene (5.1%), *a*-guainene (4.7%), and camphor (3.3%), while the Vietnamese varieties-derived essential oil contains camphor

(21.8%), germacrene D (18.3%), *a*-caryophyllene (5.6%), trans-*a*-farnesene (3.8%), and 1,8-cineole (3.1%), but no *artemisia* ketone. Woerdenbag et al.(1994) measured the maximum oil content (containing 55% of monoterpenes) prior to the flowering period in Vietnamese varieties. Hethelyi et al. (1995) analyzed Hungarian varieties-isolated essential oil content (0.48-0.81%) from the flowering shoots, and found that the essential oil mainly consists of *artemisia* ketone and *artemisia* alcohol that vary from 33% to 75% and from 15% to 56%, respectively. *A. annua* plants grown in India contains *artemisia* ketone (58.8%), camphor (15.8%), 1, 8-cineole (10.2%), and germacrene D (2.4%) (Bhakuni et al. 2001). Until now, more than 70 constituents have been investigated and identified in *A. annua* (Li et al. 2006).

Artemisinin is a potent antimalarial agent and can clean the chloroquine- and quinine-resistant *Plasmodium falciparum* strains, also known as multi-drug resistant strains. Administration of artemisinin will allow 90% of malarial patients getting rid of death within 48 h, although the disease may re-occur in a short intermittent phase. Therefore, WHO has recommended the ACTs for malaria therapy, which include other long-term antimalarial drugs in addition to artemisinin. Furthermore, artemisinin is able to kill other parasites, such as *Schistosoma japonicum*, *Clonorchis sinensis*, *Theileria annulatan* and *Toxoplasma gondii.* Although the neurotoxicity occurs in experimental animals upon high doses of artemisinin, no significant clinic toxicity appears in patients. In similar, high-dose of artemisinin induces fatal resorption in animals, but does not show any mutagenic and teratogenic effects on the pregnant women infected by severe malaria or uncomplicated malaria.

In the cytotoxicity assays of artemisinin and derivatives to Ehrlich ascite tumor cells, Woedenbag et al. (1993) estimated the 50% inhibitory concentration (IC_{50}) of artemisinin, artesunate, artemether and arteether as 12.2-29.8 μM, artemisitene as 6.8 μM, and the dimmers of dihydroartemisinin as 1.4 μM. Zheng et al. (1994) and Jung (1997) determined the significant cytotoxicity of artemisinin and semi-synthetic analogs on L-1210, P-388, A-549, HT-29, MCF-7 and KB tumor cells. Beekman et al. (1997a; 1997b) also detected the stereochemistry-dependent cytotoxicity of artemisinin and analogs.

Artemisinin even exhibits activities of anti-arrhythmia (Wang et al. 1998) and anti-hepatitis B virus (Romero et al. 2005), while artemisinic acid shows antibacterial activity (Roth and Acton 1989). Artemisinin and dihydroartemisinin play marked suppression effects on the humeral responses in mice at the high dosage, but do not alter the delayed-type hypersensitivity response to sheep erythrocytes (Tavlik et al. 1990). In a therapy trial, 56 patients with systemic lupus erythematosus (with the types of DLE 16, SCLE 10 and SLE 30) were treated with intravenous injection of artesunate (60 mg once a day, 15 days of a course, and two to four course), during which the therapy efficacies of 94%, 90% and 80% were estimated for each type of systemic lupus erythematosus. *Artemisia* ketone plays versatile roles of anti-inflammation, angiotensin converting enzyme inhibition, cytokinin-like and antitumor. Polymethoxyflavones, casticin (Yang et al. 1988), artemetin, chrysosplenetin, chrysosplenol-D and circilineol (Cubukcu et al. 1990) possess the weaker activity against *P. falciparum*. The coumarin scopoletin has anti- inflammatory activity (Huang et al. 1993) and the flavonoids fisetin and patuletin-3,7- dirhamnoside are non-peptide angiotensin converting enzyme inhibitors (Lin et al. 1994;).

Furthermore, Duke et al. (1987) and Chen et al. (1987) found that artemisinin exhibits the growth inhibitory activity to plant growth and accounted reduction of root growth in lettuce for about 50% at 33 μM of artemisinin. Bagchi et al. (1997) also observed the plant growth regulatory activity of artemisinin and artesunate. In addition, Shukla et al. (1992) described the plant growth regulatory activity of abscisic acid, abscisic acid methyl ester and bis(1-hydroxy-2-methylpropyl) phthalate. All these results demonstrated that artemisinin and other terpene metabolites presented in *A. annua* may be used as potential herbicides in agriculture.

3. ELUCIDATION OF *IN VIVO* ARTEMISININ BIOSYNTHETIC PATHWAY

3.1. Unconfirmed Artemisinin Biosynthetic Pathway Deduced from Intermediates

Although the complete pathway for artemisinin biosynthesis has not been established, most biochemical intermediates have been identified and some enzyme catalytic steps elucidated *in vitro* and *in vivo*. Akhila et al. (1987) proposed an artemisinin biosynthetic pathway that starts from mevalonate, in which inclusion of the intermediate candidates, isopentenylpyrophosphate (IPP), farnesylpyrophosphate (FPP), germacrane skeleton, dihydrocostunolide, cadinanolide and arteannuin B were suggested. Later, these authors detected generation of artemisinic acid from mevalonate (Akhila et al. 1990). In *A. annua,* artemisinic acid is 8-10 times abundant as artemisinin, so artemisinic acid was suggested as another biogenetic precursor of artemisinin (El-Feraly et al. 1986; Roth et al. 1987; Jung et al. 1990; Sangwan et al. 1993).

By the radioactive isotope-labeled precursor feeding, Wang et al. (1988) converted ^{3}H-labeled artemisinic acid (C-15) to arteannuin B and artemisinin although via separate pathways, so they tentatively concluded that artemisinic acid may be a common precursor of arteannuin B and artemisinin. This result was later supported by Sangwan et al. (1993), who confirmed *in vitro* and *in vivo* transformation of artemisinic acid to arteannuin B and artemisinin. Other confirming experimental data were then compiled by several research groups. Kudakasseril et al. (1987) and Martinez and Staba (1988) converted IPP to arteannuin B and artemisinin. Nair and Basile (1993) and Roth and Acton (1989) converted arteannuin B into artemisinin. Bharel et al. (1998) accompleshed *in vitro* biotransformation of artemisinic acid, artemisinin B and dihydroartemisinin B to artemisinin. However, in a similar experiment, Wang et al. (1993) found that artemisinin B is not a precursor to artemisinin.

On the other hand, as an endoperoxide closely related to artemisinin, artemisitene was ever isolated and characterized from *A. annua* (Acton et al. 1985). Artemisitene presents at all stages of development with amounts ranging from 0.002% to 0.09% dry mass, and the ratio of artemisitene to artemisinin increases from 1:10 in the early growth stage to 1:1 when flowers develop. Therefore, artemisitene was also reasonably considered as a candidate precursor to artemisinin. Afterwards, Kim and Kim (1992) transformed dihydroartemisinic acid into artemisinin by the cell-free extracts from teratoma cells but not from leaves or calli of *A. annua*. Li et al. (1994) synthesized [15-^{14}C]-labeled artemisinin in the supernatants prepared from the tender leaves of ripe *A. annua* with addition of [15-^{14}C] dihydroartemisinic acid as a starting compound.

Abdin et al. (2000) isolated two kinds of proteins from *A. annua* leaf cell extracts and confirmed their involvement in conversion from artemisinic acid to artemisinin. An enzyme that catalyzes artemisinin B to generate artemisinin in *A. annua* leaves was partially purified by Dingra et al. (2000) and then completely purified by Dhingra and Narasu (2001).

Unfortunately, the target genes encoding those above enzymes have not been cloned even past so many years. Following isolation of amorpha-4, 11-diene synthase (ADS), amorpha-4, 11-diene was verified as the committed product from the cyclization reaction of FPP (Bouwmeester et al. 1999). Later, Wallaart et al. (2000) found that increased artemisinin content is concomitant with decreased dihydroartemisinic acid content in *A. annua*. Meanwhile, those plants with increased artemisinin content exhibit a higher dihydroartemisinic acid level, but a lower artemisinic acid level.

This result led them to conclude that dihydroartemisinic acid rather than artemisinic acid is an immediate precursor of artemisinin, and that conversion from dihydroartemisinic acid to the corresponding hydroperoxide may represent a rate-limiting step during artemisinin biosynthesis. Bertea et al. (2005) detected from *A. annua* not only relevant intermediates including (dihydro)artemisinic alcohol, aldehyde and acid in addition to artemisinic acid, but also multiple enzymatic activities presumably derived from unidentified cytochrome P450 enzyme(s) that at least involve amorphadiene hydroxylase, artemisinic alcohol dehydrogenase and artemisinic aldehyde dehydrogenase. As expected by above presume, a multi-functional enzyme, cytochrome P450 monooxygenase (CYP71AV1), was experimentally identified by Teoh et al. (2006) from *A. annua* and further confirmed by the Keasling Laboratory through expression of the recombinant *CYP71AV1* gene in microbes (Ro et al. 2007; Chang et al. 2007).

The cytosolic isoprene or terpene metabolic pathway in plants, now known as the mevalonate (MVA) pathway that is involved in artemisinin biosynthesis, can be divided into the upstream common stage that presents in all plants and the downstream specific stage that only sequesters in *A. annua*. It has been known that the linear terpene backbone is synthesized in the upstream route from acetyl coenzyme A to FPP, and the circular terpene precursor is generated during the downstream phase from FPP to amorpha-4,11-diene, artemisinic alcohol, aldehyde, acid and artemisinin in *A. annua*. However, the exact chemical process from amorpha-4,11-diene to artemisinin acturally remains suggestive and awaits further experimental elucidation.

Based on all available data, Bertea et al. (2005) have suggested an integrative and bidirectional biosynthetic pathway from amorpha-4,11-diene to artemisinin, in which one direction is via the currently detected intermediates *in vivo*, (dihydro) artemisinic alcohol, aldehyde, and acid; another direction is with inclusion of the previously identified arteannuin B, artemisitene and artemisinic acid (Figure 4).

3.2. Possibly Concomitant or Exclusive Presence of Enzymatic and Non-Enzymatic Reactions

The possibility regarding concomitant presence of the bidirectional pathways toward artemisinin either from artemisinic acid or dihydroartemisinic acid can not be completely eliminated, but the detailed mechanism and subsequent fate (continuous conversion or as a final product?) are uncertain. Brown and Sy (2004) fed the intact *A. annua* plant with isotope-labeled dihydroartemisinic acid and detected 16 kinds of 12-carboxyamorphane and cadinane sesquiterpenes that include a small proportion of labeled artemisinin, suggesting that dihydroartemisinic acid is converted to artemisinin. Furthermore, they also confirmed that the committed product of dihydroartemisinic acid is an allylic hydroperoxide that originates from a non-enzymatic catalysis by the molecular oxygen rather than from an enzymatic step. This observation led them to conclude that the main 'metabolic route' for dihydroartemisinic acid in *A. annua* involves a spontaneous autooxidation mechanism.

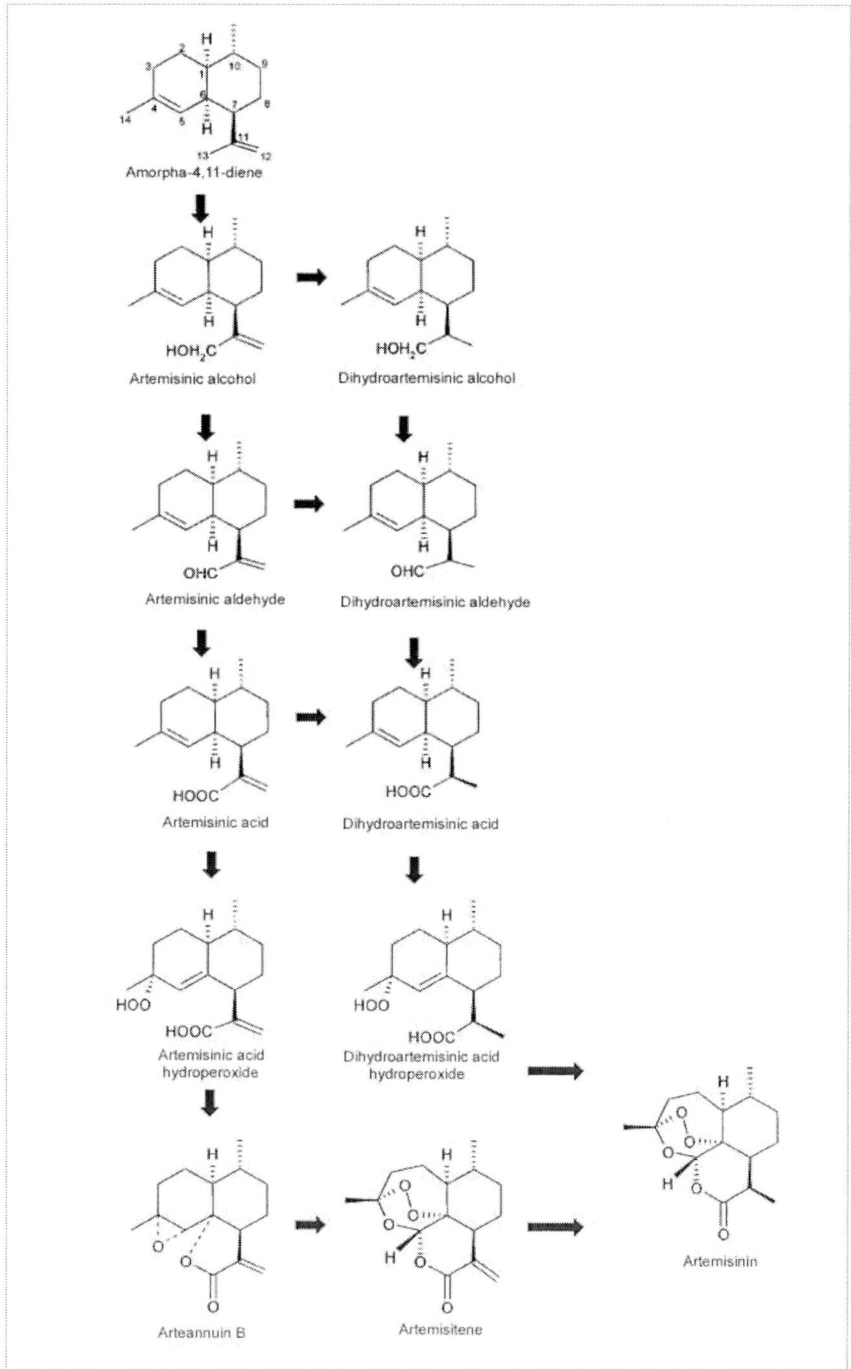

Figure 4. The proposed biosynthetic pathway from amorpha-4,11-diene to artemisinin. Some of the enzymatic or non-enzymatic steps shown by the dash arrows and intermediates indicated in the figure are not yet clearly identified. Adopted from Zeng et al. (2008a).

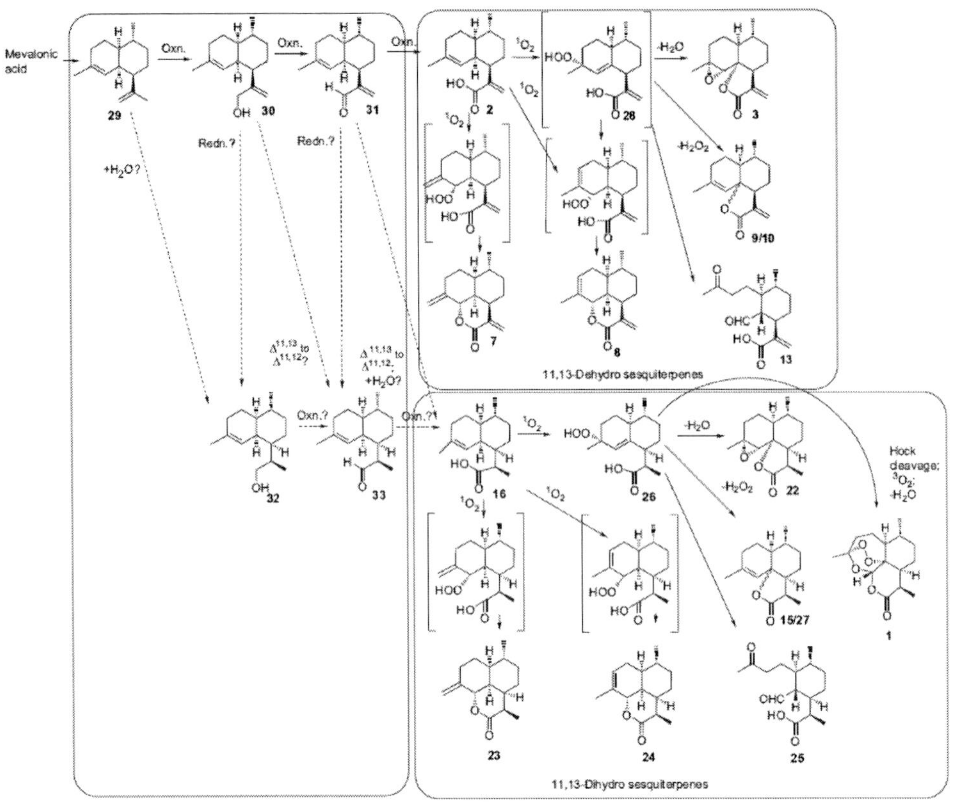

Figure 5. A unified biosynthetic pathway deduced from isotope-labeling data of 11,13-unsaturated or saturated sesquiterpenes in A. annua. 1. artemisinin; 2. artemisinic acid; 3. arteannuin B; 7. annulide; 8. isoannulide; 9. epi-deoxyarteannuin B; 10. deoxyarteannuin B; 13. seco-cadinane; 15. dihydro-epi-deoxyarteannuin B; 16. dihydroartemisinic acid; 22. dihydroarteannuin B; 23. arteannuin I; 24. arteannuin J; 25. seco-cadinane; 26. allylic hydroperoxide of dihydroartemisinic acid; 27. dihydro-deoxyarteannuin B; 28. allylic hydroperoxide of artemisinic acid; 29. amorpha-4,11-diene; 30. artemisinic alcohol; 31. artemisinic aldehyde; 32. dihydro artemisinic alcohol; 33. dihydroartemisinic aldehyde. Oxn. oxidation; Redn. Reduction. Adapted from Brown and Sy (2007).

If this conclusion is neatly reaching the reality, we are able to reasonably infer that the non-enzymatic pathway directing dihydroartemisinic acid to artemisinin is likely presented exclusively in *A. annua*. Brown and Sy (2007) fed isotope-labeled artemisinic acid to *Artemisia annua* and isolated seven labeled sesquiterpene metabolites, comprising of arteannuin B, annulide, isoannulide, *epi*-deoxyarteannuin B, deoxyarteannuin B, *seco*-cadinane and artemisinic acid methyl ester, but not including artemisinin. The fact that all of the sesquiterpenes retain their unsaturation at the 11,13-position and do not convert into any 11,13-dihydro metabolites allowed them to draw a conclusion that artemisinic acid-derived amorphane and cadinane sesquiterpenes do not convert to artemisinin (Figure 5).

Nevertheless, the authors also acknowledged that although any significant 'direct' pathway from artemisinic acid to artemisinin appear to be ruled out, but it is still possible that the 'indirect' route via arteannuin B may be presented. In addition, due to a short-term feeding course, there may be not sufficient time to accumulate a detectable amount of labeled artemisinin.

This conclusion is conceivable if the conversion rate from arteannuin B to artemisinin is much slower than the observed conversion rate from artemisinic acid to arteannuin B. Therefore, this issue may be resolved, in the future, by using the isotope-labeled arteannuin B as a feeding precursor. From such conclusive result, we are able to figure out although two parallel pathways toward artemisinic acid and dihydroartemisinic acid are concomitantly presented in *A. annua*, their subsequent fatalities are totally distinct, i.e. while dihydroartemisinic acid can convert to artemisinin, artemisinic acid cannot.

3.3. Experimental Evidence on Dihydroartemisinic Acid as a ROS Scavenger and Artemisinin as a ROS Pool

The installation of endoperoxide bridge structure in artemisinin, either enzymatically or non-enzymatically, is still not conclusive. Wallaart et al. (1999a; 1999b) found a novel intermediate dihydroartemisinic acid hydroperoxide that was previously undetected in *A. annua*, and suggested that dihydroartemisinic acid hydroperoxide may be derived from oxidation of dihydroartemisinic acid by $^1O_2$. From this point, they further proposed that dihydroartemisinic acid is first converted to dihydroartemisinic acid hydroperoxide via a light-involved and $^1O_2$-catalyzed reaction, and then the resultant dihydro- artemisinic acid hydroperoxide is autooxidized to artemisinin in air. Such deduced reaction mechanism underlying the non-enzymatic conversion from dihydroartemisinic acid to artemisinin has been previously validated by Sy and Brown (2002), who demonstrated that dihydroartemisinic acid can be slowly autooxidized into artemisinin through two steps, the first step involves light but the second step completes in the dark.

Since Knox and Dodge (1985) concluded that $^1O_2$ can be generated and emitted from the plant cell as exposure to CO_2 shortage, freezing, strong irradiation and supplement with photosynthesis inhibitors, more reports have indicated that artemisinin production is strongly influenced by the climatological conditions (Chen and Zhang 1987; Martinez and Staba 1988; Ferreira et al. 1995). The post-harvest drying process of *A. annua* is beneficial to boosting artemisinin accumulation (Laughlin 1993). Environmental stresses, such as extreme light, temperature, water and salt, significantly alter the artemisinin yield (Weathers et al. 1994). Irfan et al. (2005) reported that high concentration of salts and heavy metals augment the intracellular osmotic pressure and lead to high efficient conversion to artemisinin. Lommen et al. (2007) found that the artemisinin content in dead leaves is higher than young leaves in *A. annua,* thereby potentiating the previous postulation of post-harvest drying effects on artemisinin accumulation. Yin et al. (2008) established a positive correlation of chilling stress to overexpression of artemisinin biosynthetic genes and to overproduction of artemisinin.

It seems that dihydroartemisinic acid may act as a scavenger of ROS capable of specifically quenching $^1O_2$ in cytosol, like the carotenoids in chloroplast, for protecting the mesophyll cells from oxidative stress-mediated damage. Nevertheless, what is the real ecological implications of artemisinin and how the cytosolic dihydroartemisinic acid cooperates with the plastidic carotenoids in dealing with the harmful $^1O_2$ needs more detailed investigations.

3.4. Crosstalk and Flux Exchange between Cytosolic and Plastidic Terpene Pathways for Artemisinin Biosynthesis

It has been know that in higher plants, two independent pathways that locate in the separate intracellular compartments are involved in terpene synthesis: the cytosolic MVA pathway and the plastidic non-MVA pathway, also called 2-*C*-methyl-*D*-erythritol-4-phosphate (MEP) or 1-deoxy-*D*-xylulose-5-phosphate (DXP) pathway (Newman et al. 1999). The MVA pathway provides the biogenic precursors of sesquiterpenes, triterpenes (including sterols and brassinosteroids), polyterpenes (e.g. dolichol), polyprenols and the phytohormone cytokinin, whereas the MEP/DXP pathway is involved in generation of monoterpenes, diterpenes (e.g. phtoenol), photosynthesis-related terpenoids such as carotenoids, plastoquinone, phylloquinones and the side chains of chlorophylls, and phytohormones including abscisic acid (ABA) and gibberellins. In the cytosol, farnesyl pyrophosphate (FPP) is as a common precursor for all terpene synthesis; In plastids, geranylpyrophosphate (GPP) is a precursor of monoterpenes, while geranylgeranyl pyrophosphate (GGPP) is a precursor of diterpenes, tetraterpene and plastoquinones (Mahmoud and Croteau 2002). In addition, except for the cytosolic and plastidic pathways, there is an incomplete mitochondrial pathway that uniquely involves ubiquinone biosynthesis (Figure 6).

As seen from Figure 6, the shuttle intermediate among different compartments for carbon flux exchange is IPP, which can be transported into/out of plastids through the membrane-located transporters. Whether the IPP pool stored in *A. annua* plastids contributes a much proportion of carbon source to artemisinin biosynthesis carrying out in the cytosol is yet to be elucidated.

Although the subcellular compartmentations allow two distinct pathways to operate independently, there is increasing evidence that they well cooperate in metabolite biosynthesis. For example, sesquiterpene labeling and quantitative ^{13}C- nuclear magnetic resonance spectroscopy showed that the chamomile sesquiterpene is composed of two C5 terpenoid units formed via the MEP/DXP pathway with a third unit being derived from both the MVA and MEP/DXP pathways (Adam and Zapp 1998). In *Arabidopsis thaliana*, the MEP/DXP pathway can compensate for the reduced flux through the inhibited MVA pathway and *vice versa* (Laule et al. 2003). Using a tobacco Bright Yellow-2 cell suspension system, Hemmerlin et al. (2003) also investigated the cross-talk between such two subcellular pathways by incorporation of the labeled 1-deoxy-*D*-xylulose into the intact plants. Their results indicated that the sterols normally derived from MVA pathway can be also synthesized via the MEP/DXP pathway in presence of an inhibitor of HMG-CoA reductase (HMGR), mevinolin (MEV), and that growth inhibition caused by an inhibitor of DXP reductoisomerase (DXR), fosmidomycin (FSM), can be partially overcome by the MVA pathway (Hemmerlin et al. 2003).

Evidence of cross-talk between pathways has also been documented in other plant species (Croteau et al. 2000). For instance, interaction of both pathways on biosynthesis of monoterpenes and sesquiterpenes in lima beans (Piel et al. 1998), of gibberellins in *Arabidopsis* (Kasahara et al. 2002), and of sesquiterpene germacrene D in *Solidago Canadensis* (Steliopoulis et al. 2002). All these results indicated that MVA-derived precursors seems to be imported into the plastids, and correspondingly, MEP/DXP- derived precursors can also be exported to the cytosol. However, Dudareva et al. (2005) convinced that the MEP/DXP pathway provides IPP for both plastidic monoterpene and cytosolic sesquiterpene biosynthesis in the epidermis of snapdragon petals, but the trafficking of IPP occurs unidirectionally from plastids to the cytosol.

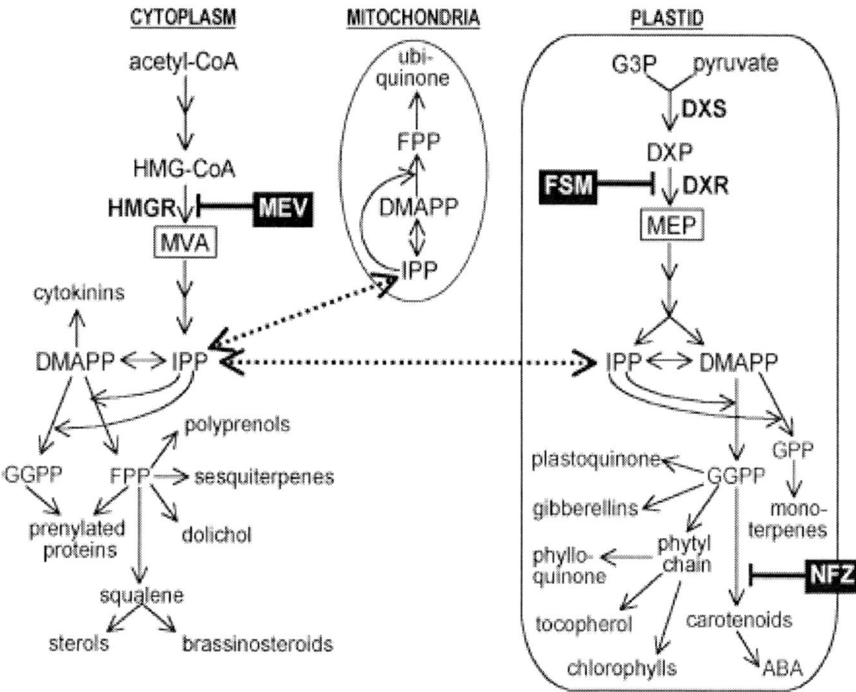

Figure 6. Isoprenoid biosynthetic pathways in distinct subsellular compartments of plant cells. ABA: abscisic acid; DMAPP: dimethylallyl diphosphate; DXP: deoxyxylulose 5-phosphate; DXR: DXP reductoisomerase; DXS: DXP synthase; FPP: farnesyl pyrophosphate; FSM: fosmidomycin; GGPP: geranylgeranyl pyrophosphate; HMG-CoA: 3-hydroxy-3-lmethylglutaryl coenzyme A; HMGR: HMG-CoA reductase; G3P: glyceraldehyde 3-phosphate; GPP: geranyl pyrophosphate; IPP: isopentenyl pyrophosphate; MEP: methylerythritol 4-phosphate; MEV: mevinolin; MVA: mevalonic acid; NFZ: norflurazon. Adopted from Rodriguez- Concepcion et al. (2004).

Recent experimental results provided much support to presence of the exporter system on the plastid envelope membranes that enable terpenoid intermediates cross out of plastids in *Arabidopsis thaliana*, spinach, kale, and Indian mustard. The isolated chloroplast envelopes can transport IPP and GPP with higher efficiencies, but transport FPP and DMAPP with lower rates. Such conclusion implied that there are specific transporters embedded in the plastid bilayer membranes (Bick and Lange 2003). Flugge et al. (2005) confirmed that plastidic phosphate translocator family members are capable of exporting the terpene intermediates across the plastid envelopes.

Despite there is ample experimental data regarding the metabolic cross-talk between two pathways, little is known about the regulation manner of flux exchange. After treating *A. thaliana* with MEV or FSM, researchers found that the dramatic inhibitor-mediated changes in the end-product levels are not reflected in the altered gene expression levels of biosynthetic enzymes (Laule et al. 2003), thereby indicating that the post- transcriptional event plays a major role in the temperal regulation of the pathways. Some novel regulatory factors comprising exoribonuclease polyribonucleotide phosphorylase (Sauret-Gueto et al. 2006) and pentatricopeptide repeat protein (Kobayashi et al. 2007) were thought to regulate the normal accumulation of enzyme quantities by a post- transcriptional regulation mode. Moreover, phytochromes may also regulate biosynthesis and exchange of terpenoid precursors through light perception and signal transduction (Rodriguez-Concepcion et al. 2004).

Adam et al. (1998) and Steliopoulis et al. (2002) showed that the sesquiterpene-directed IPP carbon comes from both the MEP/DXP and MVA pathways in chamomile and goldenrod, which are members of the Compositae family and closely related to *A. annua*. Now the question is whether the MEP/DXP pathway contributes to artemisinin biosynthesis that normally take place via the MVA pathway. By treating *A. annua* plants with the MVA pathway inhibitor MEV or/and MEP/DXP pathway inhibitor FSM, Towler and Weathers (2007) found that artemisinin production is significantly reduced in presence of each inhibitor, thereby proving that both pathways are involved in artemisinin biosynthesis. Our experimental data also demonstrated that both the pathway inhibitors, lovastatin (LVT) and FSM promote transcription of genes that encode the plastidic HMGR and DXR in *A. annua*, in which block of the MVA pathway with 5 μM of LVT led to twofold elevation of *hmgr* and *dxr* mRNA levels, and block of the MEP/DXP pathway with 100 μM of FSM led to threefold elevation of *hmgr* and *dxr* mRNA levels (data not shown). These data indicated a possible metabolic cross-talk between the cytosolic and plastidic pathways in regard of terpene biosynthesis in *A. annua*.

4. Genetic and Environmental Regulation Mechanisms on Artemisinin Biosynthesis

4.1. Artemisinic Biosynthetic Genes and Novel EST Cloning from *A. annua*

Since 1995, a dozen of genes related to artemisinin biosynthesis have been cloned from *A. annua*, and their complete or partial mRNA sequences accessed in GenBank (Table 3).

At present, functional genomic investigations regarding artemisinin biosynthesis in *A. annua* have been emphasized in several laboratories (Bertea et al. 2006; Covello et al. 2007; Zeng et al. 2008b), which would, in the near future, lead to discovery of a large batch of useful genes encoding structural and regulatory proteins involved in the anabolism of artemisinin and other valuable secondary metabolites. While two plastidic enzymes, DXS and DXR, in the MEP/DXP pathway may be indirectly responsible for artemisinin biosynthesis, other cytoplasmic enzymes including ADS, CYP71AV1 and CPR are particularly relevant to such a process.

Bouwmeester et al. (1999) first reported the partial purification of natural ADS enzyme from *A. annua*. This enzyme has a pH optimum around 6.5-7.0, a K_m of 0.6 μM for FPP, and a molecular weight of 56 kDa. The *ADS* gene were later identified and cloned by several groups independently (Mercke et al. 2000; Chang et al. 2000; Wallaart et al. 2001).

The recombinant ADS prepared from *E. coli* by Mercke et al. (2000) has a broad pH optimum between 7.5-9.0, a K_m of 0.9 μM for FPP at pH 7.5, and a molecular mass of 63.9 kDa, which is higher than the native plant ADS identified by Bouwmeester et al. (1999). Expression of *ADS* gene in *E. coli* leads to generation of a number of terpene precursors, predominantly amorpha-4,11-diene (91.2%, w/w). Picaud et al. (2005) reported that the recombinant ADS expressed in *E. coli* gives rise to amorpha-4,11-diene as a major product, but 15 sesquiterpenes in total are simultaneously produced.

Table 3. Artemisinin biosynthetic genes and other related *A. annua* genes ccessed in GenBank

Gene	Enzyme/location/sequence type	GenBank accession	Submitter
HMGR /HMG	3-hydroxy-3-methylglutaryl coenzyme A (HMG-CoA)		
	reductase, cytosol	AAU14624	Kang et al. 1995
	mRNA, complete cds	AAU14625	Kang et al. 1995
	mRNA, complete cds	AF142473	Chen et al. 1999
	mRNA, partial cds	AF156854	Chen et al. 1999
	mRNA, complete cds		
FPPS/FPS	farnesyl diphosphate (FPP) synthase, cytosol		
	mRNA, complete cds	AF136602	Chen et al. 1999
	mRNA, complete cds	AF149257	Liu et al. 1999
	mRNA, complete cds	AF112881	Chen et al. 2000
DXS	1-deoxy-D-xylulose-5-phosphate synthase (DXS),		
	plastid	AF182286	Wobbe et al. 2000
	mRNA, complete cds		
DXR	1-deoxy-D-xylulose-5-phosphate reductoisomerase		
	(DXR), plastid	AF182286	Wobbe et al. 2000
	mRNA, partial cds	AF182287	Wobbe et al. 2000
	mRNA, complete cds		
ADS	amorpha-4,11-diene synthase (ADS), cytosol		
	mRNA, complete cds	AF138959	Mercke et al. 2000
	mRNA, complete cds	AF327527	Liu et al. 2001
	mRNA, complete cds	AY006482	Wallaart et al. 2001
	mRNA, complete cds	AJ251751	Chang et al. 2005
	mRNA, complete cds	DQ241826	Huang et al. 2005
	mRNA, complete cds	EF197888	Kong et al. 2007
CYP71AV1	cytochrome P450 monooxygenase (CYP71AV1), endoplasmic reticulum		
	mRNA, complete cds	DQ268763	Ro et al. 2006
	mRNA, complete cds	DQ315671	Teoh et al. 2006
	mRNA, complete cds	DQ453967	Olsson et al. 2006
	mRNA, complete cds	DQ872632	Yin et al. 2006
	mRNA, complete cds	DQ667171	Kong et al. 2006
	mRNA, complete cds	EF197889	Kong et al. 2007
CPR	cytochrome P450 reductase (CPR), cytosol		
	mRNA, complete cds	DQ104642	Ro et al. 2006
	mRNA, complete cds	DQ318192	Yin et al. 2006
	mRNA, complete cds	EF197890	Kong et al. 2007
SS/SQS	squalene synthase (SS), endoplasmic reticulum		
	mRNA, complete cds	AF181557	Wobbe et al. 1999
	mRNA, complete cds	AF405310	Liu et al. 2001
	mRNA, complete cds	AY445505	Zeng et al. 2003

As a huge plant gene family, *CYP* gene family exhibits tremendous sequence diversity among their member genes (Schuler and Werck-Reichhart 2003). Teoh et al. (2006) isolated a cDNA clone, designated as *CYP71AV1*, from an *A. annua* glandular trichome library, and identified *CYP71AV1* gene belonging to *CYP71D* subfamily that encodes a lot of plant terpene hydroxylases. They assayed the yeast microsomal membrane-bound recombinant

CYP71AV1 with a variety of substrates in NADPH involvement. When amorpha-4,11-diene was supplemented, artemisinic alcohol synthesized in an NADPH- dependent manner. Their experiment suggested that CYP71AV1 is a multifunctional enzyme that converts amorpha-4,11-diene to artemisinic acid through three sequential oxidation steps.

Alternatively, by analyzing the data pooling in a database of expressed sequence tags (ESTs) that focuses on *Lactuca*, *Helianthus* and other Compositae plants (http://www. cgpdb. ucdavis. edu), Ro et al. (2006) identified two major *CYP* subfamilies, *CYP71* and *CYP82*. Using degenerate primers designed based on the conserved sequence of *CYP71* subfamily, a full-length cDNA that encodes an open reading frame of 495 amino acids, i.e. *CYP71AV1*, was isolated from *A. annua*. When *CYP71AV1* was co-expressed with *ADS* in *S. cerevisiae*, amorpha-4,11-diene is promptly oxidized to artemisinic acid in the engineered yeast cells.

On the other hand, Bertea et al. (2006) constructed a cDNA library starting from the total RNA isolated from the glandular trichomes of *A. annua*. About 900 of randomly selected clones were partially sequenced and analyzed for sequence homologies using the BLAST algorithm. Fragment assembly identified a total of 459 contigs and 900 ESTs and then assigned functions based on the highest similarity. The enzyme types encoded by ESTs are listed in Table 4.

Table 4. Selection of isoprene biosynthetic ESTs identified from *A. annua*

EST identification	No. of hits
deoxyxylulose 5-phosphate synthase	3
isopentenyl diphosphate isomerase	1
geranyl diphosphate synthase (small subunit)	1
pinene synthase	3
limonene synthase	2
linalool synthase	8
farnesyl diphosphate synthase	1
amorpha-4,11-diene synthase	1
germacrene A synthase	2
other sesquiterpene synthases	4

In our recent work, homology of newly isolated sequences with the accessed genes in GenBank were browsed by the online BLAST software, thereby conferring homology- based functional annotations of these sequences. Among 75 accessed *A. annua* sequences in a format of either CoreNucleotide or ESTs in GenBank, four full-length cDNAs are highly homologous to the known *A. annua* genes, other 71 ESTs do not have sequence records in *A. annua*, but in which 34 ESTs are homologous to other plant genes, including 24 known protein-coding sequences and 10 unknown protein-coding sequences, other 27 ESTs do not have sequence records in any plants. Table 5 list all *A. annua* cDNAs and ESTs with annotated function. Besides, ten sequences classified as gene fragments encoding unknown proteins homologous to other plant genes, and 37 sequences recognized as gene fragments encoding unknown proteins without homology to any plant genes are also listed.

Table 5. Functional annotation of *A. annua* cDNAs and ESTs by homology comparison

GenBank accession No.	Homology comparison-based functional annotation
AY445506	*A. annua* squalene synthase, SS
DQ241826	*A. annua* amorpha-4, 11-diene synthase, ADS
DQ838799	unknown protein without homology to any plant genes
DQ872632	*A. annua* cytochrome P450 monooxygenase, CYP
DQ984181	*A. annua* cytochrome P450 reductase, CPR
DQ838800	vacular processing enzyme-1b
DQ838801	membrane protein
EF050423	structural constituent of ribosome
EF050424	chitinase, CHI
EF050425	unknown protein without homology to any plant genes
EF050426	unknown protein without homology to any plant genes
EF050427	ribulose-1,5-bisphosphate carboxylase/oxygenase small subunit, RuBPC/O
EF050428	unknown protein without homology to any plant genes
EF050429	cytosolic NADP-malic enzyme
EF379388	unknown protein without homology to any plant genes
EF494771	unknown protein without homology to any plant genes
EF494772	unknown protein without homology to any plant genes
EF494773	unknown protein without homology to any plant genes
EF549580	40S ribosomal protein S9
EF549581	15.9kDa subunit of RNA polymerase II
EF549582	calmodulin, CaM
EF549583	histone H4-like protein
EF549584	light-harvesting chlorophyll a/b-binding protein, LHCBP
EF549585	ubiquitin-conjugating enzyme, UCE
EF660343	eukaryotic translational factor TIF3B1
ES494773	unknown protein homologous to other plant genes
ES582125	drought/low temperature and salt responsive protein, D/LTSRP
ES582126	unknown protein homologous to other plant genes
ES582127	unknown protein without homology to any plant genes
ES582128	hydroxyl praline-rich protein
ES582129	RNA-binding glycine-rich protein, RGP
ES582130	unknown protein homologous to other plant genes
ES582131	thioredoxin
ES582132	acyl-ACP thioesterase FATA1
ES582133	unknown protein homologous to other plant genes
ES582134	unknown protein homologous to other plant genes

Table 5. (Continued)

GenBank accession No.	Homology comparison-based functional annotation
ES582135	unknown protein homologous to other plant genes
ES582136	unknown protein without homology to any plant genes
ES582137	unknown protein without homology to any plant genes
ES582138	unknown protein without homology to any plant genes
ES582139	unknown protein without homology to any plant genes
ES582140	unknown protein homologous to other plant genes
ES582141	unknown protein without homology to any plant genes
ES582142	unknown protein homologous to other plant genes
ES582143	DICER-like 2/3 spliceform 2
ES582144	unknown protein homologous to other plant genes
ES582145	auxin-repressed/dormancy-associated protein, AR/DAP
ES582146	unknown protein without homology to any plant genes
ES582147	unknown protein without homology to any plant genes
ES582148	unknown protein without homology to any plant genes
ES582149	unknown protein without homology to any plant genes
ES582150	unknown protein without homology to any plant genes
ES582151	unknown protein homologous to other plant genes
ES582152	cytosolic malate dehydrogenase
ES582153	40S ribosomal protein S30-like protein
ES582154	secretory peroxidase
ES582155	unknown protein homologous to other plant genes
ES880929	unknown protein without homology to any plant genes
ES880930	unknown protein without homology to any plant genes
ES880931	unknown protein without homology to any plant genes
ES880932	unknown protein without homology to any plant genes
ES880933	unknown protein without homology to any plant genes
ES880934	unknown protein without homology to any plant genes
ES880935	unknown protein without homology to any plant genes
ES880936	unknown protein without homology to any plant genes
ES880937	unknown protein without homology to any plant genes
ES880938	unknown protein without homology to any plant genes
ES880939	unknown protein without homology to any plant genes
EV780877	unknown protein without homology to any plant genes
EV780878	unknown protein without homology to any plant genes
EV780879	unknown protein without homology to any plant genes
EV780880	unknown protein without homology to any plant genes
EV780881	unknown protein without homology to any plant genes
EV780882	unknown protein without homology to any plant genes
EV780883	unknown protein without homology to any plant genes

From the known functions served by above *A. annua* genes, one can find out a lot of genes coding for the environment inducible or stress responsive proteins. The diverse genes include those for metabolic engineering-directed breeding for boosting artemisinin production (*ADS*, *CYP71AV1*, *CPR*, and *SS*), those for breeding toward the highly effective photosynthesis (*RuBPC/O*, and *LHCBP*, etc.), and those for breeding on the disease and pest insect resistance (*CHI*, *D/LTSRP*, *AR/DAP*, and *CaM*, etc.). Moreover, all novel genes with unknown protein-encoding function or no sequencing records await further identifications by the so-called 'gain/lost-of-function' approaches, such as site-directed gene mutagenesis, homologous recombination-mediated gene knockout, and anti-sense inhibition or microRNA interference.

4.2. Temporal and Spatial Expression Patterns of Artemisinin Biosynthetic Genes

Teoh et al. (2006) proved that *CYP71AV1* gene is expressed in *A. annua* at a maximum level in glandular trichomes, a moderate level in leaves, and a minimum level in roots, which is just in accordance with the distribution pattern of artemisinin, i.e. glandular trichomes give rise to the highest artemisinin content, leaves display a lower level of artemisinin yield, and artemisinin is presented in trace amount in roots. Our recent immunoquantitative assay of organ-specific distribution of CYP71AV1 showed that abundance of the tested enzym is highest in leaves, moderate in stems and lowest in roots (unpublished data). From artemisinin determination results throughout the vegetative stage of *A. annua*, it was known that top leaves (later initiated) exhibit generally higher artemisinin content, but bottom leaves (earlier initiated) show lower artemisinin content (Liersch et al. 1986; Ferreira et al. 1995). The previous follow-up determination of artemisinin showed that a highest artemisinin content was often achieved just before flowering (Morales et al. 1993) or under flowering (Gupta et al. 2002; Laughlin 1995). However, the highest artemisinin content was determined in dry leaves exposed to the post-harvest maturation (Lommen et al. 2006) and even in dead leaves experienced the programmed cell death (apoptosis) (Lommen et al. 2007).

Weathers et al. (2006) cited their unpublished experimental results that *ADS* mRNA is ubiquitously present in all tissues including roots, stems, leaves, and flowers in mature *A. annua* plants. However, considering the fact that artemisinin is only accumulated in the glandular trichomes on leaves and flowers, they explained that *ADS* mRNA may not be translated into protein in all tissues (translationally regulated) or ADS may not be active in all tissues (post-translationally regulated), and addressed that the most probable situation is that amorpha-4,11-diene may be synthesized in all tissues, but then transported to leaves and flowers for further artemisinin biosynthesis, or amorpha-4,11- diene synthesized in roots and stems may be used to produce compounds other than artemisinin. These suggestions need verification by quantitative assay of *ADS* mRNA in different tissues and *in situ* immunoquantitative determination of ADS concentration or enzymatic detection of ADS activity in all tissues.

4.3. Quantification on Environmental Stress-Induced Overexpression of Artemisinin Biosynthetic Genes

It is obviously from the artemisinin biosynthetic pathway that dihydroartemisinic acid as a precursor for artemisinin biosynthesis experiences the hydroperoxide intermediate stage. This means that artemisinin biosynthesis may involve induction of artemisinin biosynthetic genes evoked by environmental stresses. However, current investigations regarding stress-induced gene expression in *A. annua* were restricted to a few stress factors, e.g. extensive light illumination (Souret et al. 2002; 2003). Yin et al. (2008) treated *in vitro* cultural *A. annua* plants by cold, heat and ultraviolet light, and then quantified the transcripts of three artemisinin biosynthetic genes, *ADS*, *CYP71AV1* and *CPR*, which demonstrated that *ADS* and *CYP71AV1* genes are markedly up-regulated, while *CPR* gene keeps stable expression either prior to or post the treatment. The real-time fluorescent quantification data further revealed that as exposure to chilling stress, the copy numbers of *ADS* and *CYP71AV1* mRNAs in *A. annua* plants were accounted as eleven- and sevenfold elevations as the control plants, respectively. Nevertheless, artemisinin content in those *A. annua* plants exposured to chilling stress does not increase in proportion with elevated levels of *ADS* and *CYP71AV1* mRNAs, implying that artemisinin biosynthesis may be modulated by more than one step regulations, which determine the conversion efficiency from artemisinic acid to artemisinin.

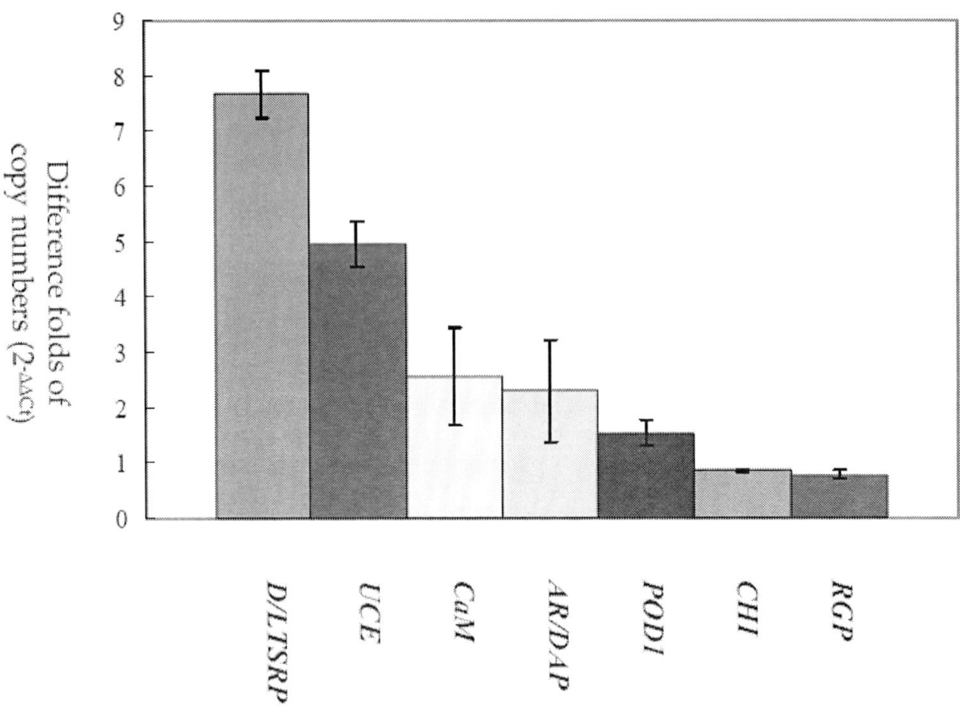

Figure 7. Regulation of chilling stress-induced expression in A. annua. AR/DAP: Auxin-repressed/dormancy- associated protein; CaM: calmodulin gene; CHI: chitinase gene; D/LTSRP: drought/low temperature and salt responsive protein gene; POD1: peroxidase 1 gene; RGP: RNA-binding glycine-rich protein gene; UCE: ubiquitin-conjugating enzyme gene. Adopted from Zeng et al. (2008b).

Wallaart et al. (1999) ascertained that only those *A. annua* shoots treated by drought (30% relative humidity) and light stress (6000 lux) can be amplified from *ADS* mRNA, but no product is available from amplification of non-treated *A. annua* shoots. In similar, as we detected the amplification products by gel electrophoresis, the amplicons from *ADS* and *CYP71AV1* mRNAs were only available from the chilling stress-exposed *A. annua* plants, while no amplicon was detected from the untreated *A. annua* plants (Yin et al. 2008). The results let us to envisage that expression of *ADS* and *CYP71AV1* genes in *A. annua* may be controlled by environmental stresses, at least by low temperature in this situation.

To experimentally verify the function of our newly isolated *A. annua* ESTs, the responsive pattern of seven novel ESTs to low temperature was chosen as an evaluation criteron during quantitatively evaluating their chilling stress-induced overexpression levels by the real-time fluorescent quantitative polymerase chain reaction. The result showed that upon standing at 4°C for 48 h, the expression levels of five ESTs (*D/LTSRP*, *UCE*, *CaM*, *AR/DAP*, and *POD1*) were significantly up-regulated, while those of other two ESTs (*CHI* and *RGP*) were not predominantly fluctuated (Figure 7).

4.4. Artemisinin Biosynthetic Gene Transcription-Based Identification of Chilling Stress Signal Transduction

It has been known that low temperature-induced gene expression is essentially involved in the signal transduction pathway. Therefore, we treated cultural *A. annua* plants with Ca^{2+} channel inhibitor $LaCl_3$ and Ca^{2+} chelator ethylene glycol tetra-acetic acid (EGTA) to assess their effects on chilling-mediated signal transduction (Zeng et al. 2008b). When supplemented with $LaCl_3$ or EGTA, chilling-induced expression of *ADS* and *CYP71AV1* genes in *A. annua* was suppressed to defferent degrees, in which the former one is more potently attanuated than the later one. As $LaCl_3$ or EGTA was depleted, chilling-induced expression of *CYP71AV1* gene recovered immediately, while that of *ADS* gene retrieved more slowly. In contrast, either with or without $LaCl_3$ or EGTA, expression of *CPR* gene was not affected by the treatment. Moreover, calmodulin gene (*CaM*) was up-regulated by 2.5 folds upon chilling exposure. Lin et al. (2004) also observed the elevated CaM content and enhanced antioxidant enzyme activity in *Populus tomentosa* during cold- acclimation. We thus inferred that cyclization from FPP to form amorpha-4,11-diene might be regulated at the transcription level, seemingly involving activation binding of transcription factor(s) with *ADS* and *CYP71AV1* promoters through the Ca^{2+}-CaM signal transduction pathway. At present, there has no report demonstrating interaction of the specific transcription factor(s) with these *A. annua* promoters. Nevertheless, the promoter sequence of *ADS* gene has been accessed in GenBank, which should earge investigation of the stress-inducible binding of transcription factors to the promoter. Wang et al. (2001; 2002) discovered that the oligosaccharide elicitor derived from *Colletotrichum* sp. can trigger the signaling cascade involving rapid Ca^{2+} accumulation, plasma membrane NAD(P)H oxidase activation and ROS release.

4.5. Stress-Responsive and Other *Cis*-Regulatory Elements in ADS Promoter

The isoprene intermediate FPP is a common precursor of sterols and sesquiterpenes. The enzymes that initiate these two branching biosynthetic pathways, squalene synthase and sesquiterpene synthases, compete for FPP. In plants, different structural types of sesquiterpenes are synthesized from FPP upon catalysis by distinct sesquiterpene synthases. At least four sesquiterpene synthases have been identified in *A. annua* (Figure 8).

What is the regulation mechanism to normally cooperate with expression of all these sesquiterpene synthase gene? When are they expressed to serve the diversed cellular function? We still do not have ideas to account for them. Although the cDNA encoding ADS was isolated by cDNA library screening (Merck et al. 2000) and degenerated reverse transcription-polymerase chain reaction (Wallaart et al. 2001), its regulated expression manner is uncertain. The indirect evidence regarding regulation of *ADS* gene expression only came from amplification of *ADS* mRNA, in which no amplicon was visible on the gel when RNA was amplified from non-stressed *A. annua* plants. This is consistent with the characteristics of terpene synthases themselves, of course, which occur in very low intracellular concentrations in plant tissues.

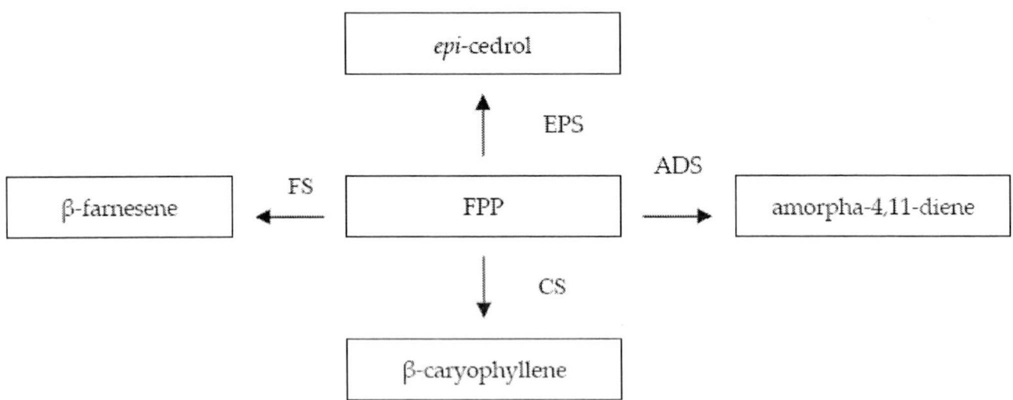

Figure 8. Sesquiterpene synthases isolated from A. annua. ADS: amorpha- 4,11-diene synthase; CS: β-caryophyllene synthase; EPS: epi-cedrol synthase; FPP: farnesyl perophosphate; FS: β-farnesene. Adopted from Weathers et al. (2006).

However, specific transcription was exclusively induced in plants pre-exposed to drought (30% relative humidity) and photo (6000 lx) stresses (Wallaart et al. 2001). The transcript level of *ADS* gene was also found to be elevated by cold, heat and ultraviolet light exposure of *A. annua* plants (Yin et al. 2008). The real-time quantitative assay showed that *A. annua* plants under a transient chilling exposure resulted in elevenfold increase in *ADS* transcript level. These results indicated that *ADS* gene may be an inducible one capable of highly responsing to stresses. Therefore, a hypothesis has been proposed that dihydroartemisinic acid may act as an antioxidant by quenching $^1O_2$ (Wallaart et al. 1999a) and generated dihydroartemisinic acid hydroperoxide consequently converted to artemisinin in response to $^1O_2$ emission (Wallaart et al. 1999b).

Several groups independently isolated and sequenced *ADS* promoter with the accessed GenBank accession No: DQ448294, DQ448295, DQ448296, DQ448297 and AY528931. Our online sequence analysis (http://www.dna.affrc.go.jp/PLACE/signalscan.html) revealed that there are a lot of potential *cis*-regulatory elements in *ADS* promoter that responses to drought, cold, salt, heat shock, mechanic wound, pathogens and phytohormones (Table 6).

Table 6. The cis-element of *ADS* promoter in *A. annua*

Cis-element	Induction	Core sequence	Location (Strand)	Reference
ABRELATERD1	dehydration	ACGTG	1064(-), 1928(-)	Simpson et al. 2003
ABRERATCAL	Ca^{2+}	MACGYGB M=C/A Y=C/T B=T/C/G	2159(+), 1415(-), 1927(-)	Kaplan et al. 2006
CRT/DRE/CBF2	cold	GTCGAC	1141(+), 1141(-)	Xue 2003
CBFHv	cold	RYCGAC R=A/G Y=C/T	59(-)	Xue 2002; Svensson et al. 2006
CCAATbox1	heat shock	CCAAT	251(+), 806(+), 844(+), 1109(+), 1507(-), 1860(-), 2068(-)	Rieping et al. 1992
GT1CONSENSUS	light	GRWAAW R=A/G W=A/T	663(+), 746(+), 752(+), 1264(+), 1317(+), 1402(+), 1488(+), 1489(+), 2230(+), 2464(+), 148(-), 283(-),694(-), 775(-), 1291(-), 1507(-), 1902(-), 149(-), 584(-), 1047(-), 2216(-), 2238(-), 2305(-)	Terzaghi et al. 1995
GT1core	light	GGTTAA	640(+), 1676(+), 642(-)	Terzaghi et al. 1995
I-box	light	GATAA	1402(+), 284(-), 585(-), 1598(-), 1903(-)	Terzaghi et al. 1995
GT1GMSCAM4	salt and pathogen	GAAAAA	663(+), 752(+), 1489(+)	Park et al. 2004
W-boxATNPR1	salicylic acid and pathogen	TTGAC	799(+), 2146(+), 316(-), 1503(-), 2295(-)	Yu et al. 2001; Chen et al. 2002
WRKY71OS	gibberellins, abscisic acid and pathogen	TGAC	244(+), 381(+), 800(+), 848(+), 941(+), 1122(+), 1195(+), 1250(+), 1666(+), 1955(+), 1977(+), 2147(+), 170(-), 316(-), 688(-), 1503(-), 1755(-), 1996(-), 2295(-)	Zhang et al. 2004; Xie et al. 2005
W-boxHVISO1	sugar	TGACT	244(+), 1195(+), 1955(+), 169(-), 315(-), 687(-), 1995(-)	Sun et al. 2003
W-boxNTERF3	wound	TGACY Y=C/T	244(+), 848(+), 941 (+), 1195(+), 1955(+), 2147 (+), 169(-), 315(-), 687(-), 1995 (-), 1502 (-), 1754 (-)	Nishiuchi et al. 2004

The future work should be identification of these *cis*-element-bound transcription factors and their *trans*-action upon environmental stresses. In this regard, the comparative genomics, transcriptomics, proteomics and metabolomics among plants with the closed evolutionary relations would be helpful and reasonable. Furthermore, direct isolation of the stress-corresponding transcription factors with the gel retardation assay and DNA fingerprinting.

5. Primary and Secondary Stress Signal-Induced Artemisinin Overproduction

5.1. Primary Stress Signal Induction

Liu et al. (1997) supplemented cell homogenate of *Aspergillus oryzae* to the cultural hairy roots of *A. annua* and detected increase of artemisinin content up to 550 mg/L. Wang et al. (2000) investigated the stimulating effects of fungal elicitors on artemisinin biosynthesis in *A. annua* hairy root cells. After adding chitosan (150 mg/L) to *A. annua* hairy root cells and keeping for six days, Putalun et al. (2007) determined artemisinin content for sixfold increase to 1.8 μg/mg dry weight, and observed a similar effect as yeast extracts (2 mg/mL) were supplemented instead. Kapoor et al. (2007) applied phosphate fertilizer to field-grown *A. annua* plants with two fungi, *Glomus macrocarpum* and *Glomus fasciculatum*, and quantified the tremendously elevated artemisinin level together with the significant high density of glandular trichomes on *A. annua* leaves.

In cultural cells and cell-free extracts of *A. annua*, supplement with sterol inhibitors, miconazol and naphtiphine, promotes incorporation of a large quantity of ^{14}C-labeled IPP into artemisinin (Kudakasseril et al. 1987). In tobacco suspension cells, supplement of a fungal elicitor leads a sharp dropped activity level of squalene synthase accompanying with a large magnitude of enhancement of sesquiterpene phytoalexin biosynthesis (Vogeli et al. 1998), which demonstrated that specific induction that is mediated by the fungal elicitors may be undergone by 'influx' into the sesquiterpene biosynthetic branch and 'efflux' from sterol biosynthetic branch.

In addition, the types of monosaccharide and disaccharide (Wang et al. 2007) and microelement deficiency (Ferreira et al. 2007) can affect artemisinin biosynthesis at a certain extent upon stress signal transduction.

5.2. Secondary Stress Signal Induction

Salicylic acid (SA), jasmonic acid (JA) and methyl jasmonate (MJ) are ubiquitous signal molecules in plant cells and known as the 'second messengers', which can transduce most of the extreme environmental stimuli to initiate the cellular responsive mechanics against the oxidative stresses. It was previously reported that MJ is a senescence- promoting inducer in *Artemisia absinthium* (Ueda et al. 1980), and MJ emitted by *Artemisia tridentate* can induce approximated tomato plants to express proteinase inhibitor gene and exhibit protection reaction (Farmer et al. 1990). The recent investigation by Afitlhile et al. (2005) showed that the plants belonging to *Artemisia* accumulate higher levels of JA and MJ than other plants, and accumulated JA is higher up to eightfold in field-grown *Artemisia* plants than in greenhouse-cultivated *Artemisia* plants, which can be divided into the high level group (30 nmol/g dry weight), moderate level group (10-20 nmol/g dry weight) and low level group (10 nmol/g dry weight). Wallaart et al. (2000) also measured various content levels in artemisinin and its precursors in *A. annua* plants, thus suggesting that different chemotypes may be mixed in *A. annua.* Nevertheless, it is unknown if these chemotypes directly lead to variation of JA and MJ levels in plants. In addition, Lommen et al. (2007) determined highest artemisinin content in senescent and dead leaves, but whether this finding is indeed related to MJ-promoted senescence and MJ-induced artemisinin overproduction needs elucidations.

Signal molecules that induce accumulation of plant secondary metabolites have more reports (Woerdenbag et al. 1993; O'Donnell et al. 1996; Walker et al. 2002). Recently, Baldi et al. (2007) evaluated the effects of SA, MJ, gibberellic acid (GA_3) and $CaCl_2$ on artemisinin biosynthesis in *A. annua* cultural cells. Consequently, they found that 20-50 mg/L SA, 5-10 mg/L MJ, and 10 mg/L GA_3 can significantly increase artemisinin content, the highest magnitude of such increase is sixfold as the control. The previous research also showed that GA_3-treated *A. annua* plants can significantly increase their artemisinin content (Fulzele et al. 1995; Paniego and Giulietti 1996). Zhang et al. (2005) demonstrated that exogenous GA_3 can divert the carbon flux to artemisinin by feed-back inhibition of GA_3 biosynthesis, thereby suggesting presence of a regulon at the conversion step from artemisinic acid to artemisinin.

To investigate the relationship between artemisinin accumulation and blooming, Wang et al. (2004; 2007) transformed *A. thaliana* flowering-promoting factor 1 gene (*FPF1*) and flowering gene (*CO*) into *A. annua.* Although transgenic plants bloom early for 20 days and 14 day, respectively, their artemisinin content does not increase significantly. Therefore, they concluded that flowering is not a prerequisite for enhanced artemisinin production in general. Weathers et al. (2005) found that *A. annua* hairy roots in the cultural medium supplemented with cytokinin 2-isoprenyl adenine give rise to substantially increased artemisinin content. After introducing *ipt* gene into *A. annua*, Geng et al. (2001) detected two- to threefold elevated cytokinin level and 30-70% boosted artemisinin content.

5.3. Combined Applications of Primary and Secondary Stress Signals

Nitric oxide (NO) is a novel signal molecule (Delledonne et al. 1998) that enhance production of taxol in *Taxus* (Wang et al. 2004) and of catharanthine in *Catharanthus* (Xu et al. 2005), but not of artemisinin in *Artemisia* (Zheng et al. 2007). However, NO can potentiate fungal elicitor-induced overproduction of ginseng saponin (Hu et al. 2003), taxol (Xu et al. 2004), hypericin (Xu et al. 2005), puerarin (Xu et al. 2006), and artemisinin (Zheng et al. 2007). These investigations also indicated that NO-mediated fungal elicitor induction of secondary metabolites is an essential outcome from the burst of ROS *en route* in JA-dependent signal transduction pathway. Zheng et al. (2007) demonstrated that artemisinin content in 20 day-old hairy roots elevates from 7 mg/g dry weight to 13 mg/g dry weight upon application of the oligosaccharide elicitor for four days, and the combined treatments of the oligosaccharide elicitor with sodium nitroprusside (SNP) lead to increase of artemisinin content even higher from 12-22 mg/g dry weight.

6. Artificial Metabolic Manipulation: Artemisinin Biosynthetic Gene Transfer into *A. annua* and Other Plants

6.1. Upstream Pathway Gene Transfer

Chen et al. (1999) employed *Agrobacterium rhizogenes* to produce transgenic *A. annua* hairy roots that express FPP synthase gene (*FPS*) driven by 35S promoter of cauliflower mosaic virus (CaMV). A number of resulting roots produce artemisinin up to 2-3 mg/g dry weight, 3-4 times that of the control roots. When *A. tumefaciens* was used instead, regenerated transgenic plants yield artemisinin at 8–10 mg/g dry weight (Chen et al. 2000). Both studies showed that although manipulation of *FPS* increase artemisinin content, the yields are only comparable to or slightly higher than wild-type plants, suggesting that the multiple downstream pathways toward so many kinds of desired or unexpected products may make it complicated in obtaining a singular target metabolite.

Davidovich-Rikanati et al. (2007) modified the flavor and aroma of tomatoes by overexpressing the *Ocimum basilicum* geraniol synthase gene under the control of the tomato ripening–specific promoter from polygalacturonase gene. A majority of untrained taste panelists preferred the transgenic fruits over controls. Monoterpene accumulation was at the expense of reduced lycopene accumulation. Similar approaches may be applicable for carotenoid-accumulating fruits and flowers in other plant species.

6.2. Antisense-Based Genetic Modification

Wang et al. (2001) identified a cytochrome P450 hydroxylase gene (CYPH) specific to the glandular trichomes and used both antisense and sense co-suppression strategies to investigate impact of such genetic modification on the plant behavior. As a result, CYPH-suppressed transgenic tobacco plants demonstrate a ≥41% decrease in the predominant exudate component, cembratriene-diol (CBT-diol), and a ≥19-fold increase in its precursor, cembratriene-ol (CBT-ol). Thus, CBT-ol level is raised from 0.2 to ≥4.3% of leaf dry weight. Exudates from antisense-expressing plants exhibit higher aphidicidal activity, and transgenic plants with exudates containing high concentration of CBT-ol show a greatly diminished aphid colonization response. Their results demonstrated the feasibility of modifying the natural product composition and aphid-interactive properties of gland exudates using metabolic engineering.

We recently introduced *A. annua* anti-sense squalene synthase gene (*asSS)* into *A. annua* to attempt enhancing artemisinin production by suppressing sterol biosynthesis. The elevated *SS* mRNA level and dropped sterol content in transgenic plants were quantified. The determination results of artemisinin content showed that in transgenic plants artemisinin yield reaches 1.66 mg/g dry weight (T47) and 1.26 mg/g dry weight (T81), while in untreated transgenic plants artemisinin content is 1.23 mg/g dry weight (T47) and 0.93 mg/g dry weight (T81). As comparison, artemisinin content in the control plant (WT) is only 4.5 mg/g dry weight.

6.3. Nuclei-Coded and Plastid-Targeted Artemisinin Precursor Biosynthetic Enzymes

Wu et al. (2006) engineered high level terpene production in tobacco plants by diverting carbon flow from cytosolic or plastidic IPP through expression of an avian *FPS* gene in the cytosol but targeting to plastids. The strategy used in the present study increased amorpha-4,11-diene content more than 1000-fold up to 25 μg/g fresh weight (Figure 9), and seems to be suitable generating high levels of other cytosolic or plastidic terpenes for scientific research, industrial production or therapeutic applications.

Terpenes represent over a $1-billion market value to the flavor and fragrance industry, but the market for pharmaceutics and medicine industry may be astonishingly attractive. Whereas tremendous progress has been made in engineering microbial platforms for terpene production, plant systems also addressed due to inexpensive carbon feedstocks, low processing costs and a greater elaboration potential. This is especially the case for those terpenes requiring decoration with carbohydrate and/or aryl substituents, and introduction of peroxide functionalities.

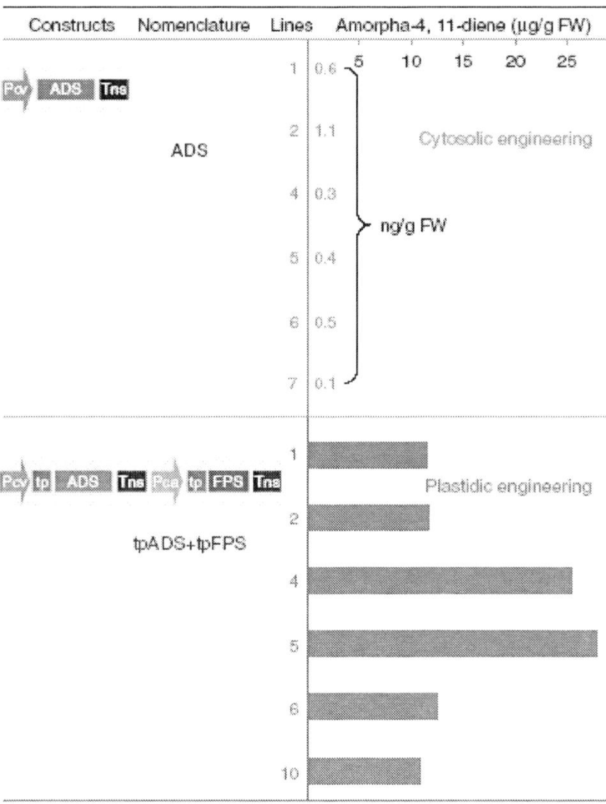

Figure 9. Engineering production platforms for amorpha-4,11-diene by diverting carbon flux from the cytosolic MVA pathway and the plastidic MEP/DXP pathway. ADS: amorpha-4,11-diene synthase; FPS: farnesyl phosphate synthase; FW: fresh weight; Pca, 35S cauliflower mosaic viral promoter; Pcv, cassava mosaic viral promoter; Tns: nopaline synthase gene (NOS) terminator; tp: plastid targeting signal sequence. Adopted from Wu et al. (2006).

7. Artificial Metabolic Manipulation: Re-Establishment of Artemisinin Biosynthetic Pathway in Microorganisms

7.1. Production of Artemisinin Precursors by Genetically Modified Yeast

As a species of eukaryotic microbes, yeast offers promise for overexpression of plant genes, especially those encoding the protein containing membrane-bound domains (DeJong et al. 2006). One of the pioneer examples is heterologous expression of *Taxus* taxane-10β-hydroxylase gene in yeast (Schoendorf et al. 2001). Another successful attempt is functional expression of *A. annua epi*-cedrol synthase gene (*ECS*) in yeast (Jackson et al. 2003). Adequate attachment of the membrane-anchored cytochrome P450 is crucial for high-level production of artemisinin in yeast. Moreover, the high rate of isoprene metabolism in yeast must be, of course, taken into account for which an ideally engineered yeast strain should be one with an increased capacity in FPP production and a decreased level of sterol accumulation.

The Brodelius group compared the effects of two different formats of *A. annua ADS* gene expression in yeast on amorpha-4,11-diene production. For doing this, *ADS* gene was either introduced into yeast as episomal plasmids or inserted into the yeast genome by homologous recombination. The plasmid-harboring and genome-integrated yeast strains with the functionally expressed *ADS* gene exhibit sixfold increase of amorpha-4, 11-diene (600 μg/L versus 100 μg/L) in 16-day batch cultivations (Lindahl et al. 2006), demonstrating that sesquiterpene production in yeast may be positively correlated with the gene dosage although the insufficient FPP pool is also a possible limiting factor.

Production of artemisinic acid in the engineered yeast strain was achieved by multiple gene transfer as illustrated in Figure 10. Firstly, *HMGR* and *ADS* genes were transferred into yeast for increasing the FPP pool and catalyzing amorpha-4,11-diene production, respectively. Further augmented levels of FPP and amorpha-4,11-diene were realized through down-regulation of *SS* gene by a methionine-repressible promoter. Subsequently, *CYP71AV1* and *CPR* genes were introduced into the engineered yeast strain to enable the three-step oxidation of amorpha-4,11-diene to generate artemisinic acid.

When evaluated step-by-step, expression of *ADS* gene alone only resulted in a small quantity of amorpha-4,11-diene (4.4 mg/L), while co-expression of a truncated *HMGR12* gene (*tHMGR*) improved amorpha-4,11-diene yield by about fivefold. Introduction of a methionine repressible promoter down-regulated the sterol biosynthetic gene, *ERG9*, and led to an additional twofold increase of amorpha-4,11-diene. Down-regulation of *ERG9* gene and overexpression of *upc2-1*, a dominant mutated gene encoding the Upc2p transcription factor capable of regulating sterol biosynthesis, elevated amorpha-4,11-diene level to 105 mg/L. Integration of an additional copy of *HMGR* gene into the chromosome enhanced amorpha-4, 11-diene production to 149 mg/L. Overexpression of *FPPS* (*ERG20*) gene decreased cell density and, in turn, increased amorpha-4,11-diene content by about additional 10% (w/w). All these multi-gene manipulations consequently led to a total enhancement in amorpha-4,11-diene output up to 153 mg/L, an elevation of nearly 500-fold of what was shown previously (Jackson et al. 2003).

In the next step, two genes, *CYP71AV1* and *CPR*, directly responsible for artemisinin biosynthesis were introduced into the engineered yeast cells to make it possible for conversion from amorpha-4,11-diene to artemisinic acid. In a shake flask, approximately 32 mg/L of artemisinic acid was measured, but only negligible artemisinic alcohol and artemisinic aldehyde were detected.

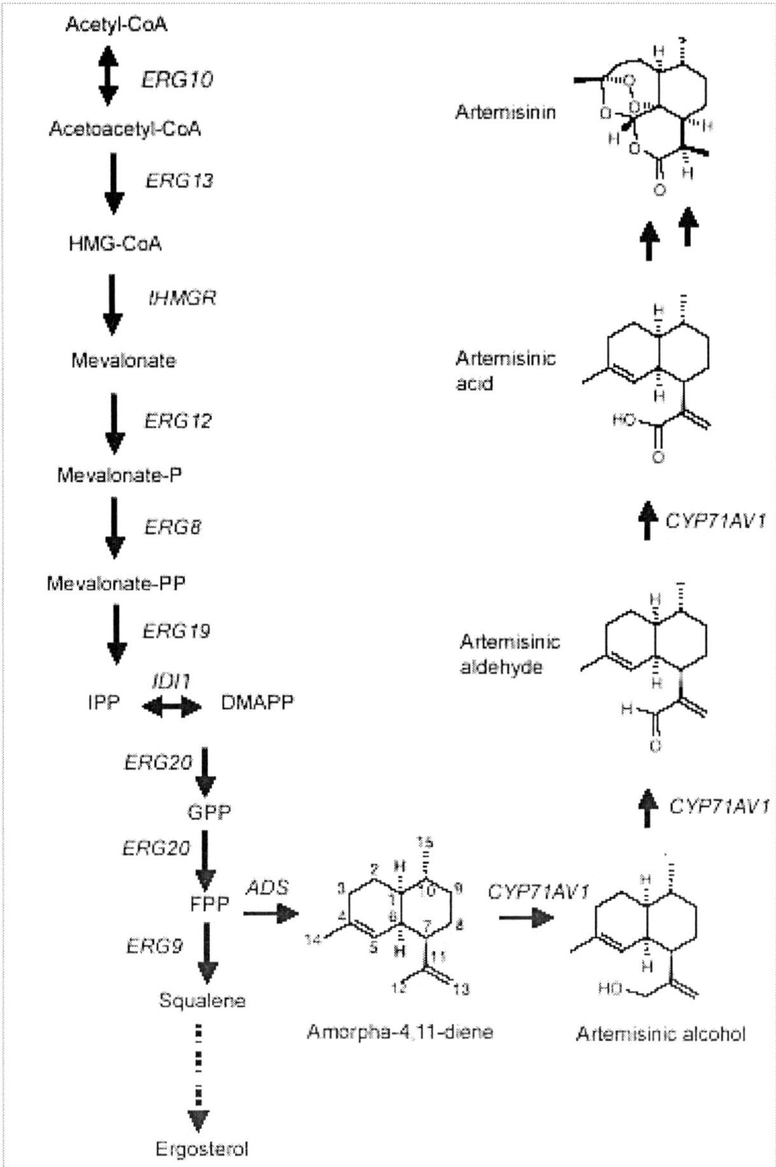

Figure 10. Engineered artemisinic acid biosynthetic pathway in S. cerevisiae. ERG: yeast MVA pathway genes; ADS: A. annua amorpha-4,11-diene synthase gene; A. annua CYP71AV1: cytochrome P450 monooxygenase gene; IDI: IPP isomerase gene; tHMGR: truncated HMG-CoA reductase gene. A. annua CPR gene is not indicated. Adopted from Zeng et al. (2008a).

The engineered yeast cells expressing *ECS* gene converted their endogenous FPP to *epi*-cedrol up to 90 μg/L. Further optimization by introducing *upc2-1* gene increased the availability of FPP and yielded 370 μg/L of *epi*-cedrol, a sixfold increase compared with what were reported in *E. coli* (Cane et al. 1993; Martin et al. 2001). The engineered yeast cells can produce comparable levels of artemisinic acid as *A. annua* on a biomass basis but in a much shorter time (4-5 days for yeast versus several months for *A. annua*). As such viewpoint, artemisinic acid productivity of the engineered yeast is nearly two orders of magnitude greater than *A. annua*, demonstrating an advantage in yeast over in plants for large-scale production of plant terpenes. Moreover, secretion of artemisinic acid to the medium by yeast cells simplifies fractionation and purification of the desired product, representing another potential cost-saving benefit in industrial processing.

7.2. Production of Artemisinin Precursors by Genetically Modified Bacteria

The classical bacterial species, *E. coli*, remains an amenable host for industrial fermentative production of plant secondary metabolites due to its speedy growth rate and easy genetic manipulation. However, difficultis in assembly of cytochrome P450 enzymes and other membrane-bound proteins renders employment of such prokaryotic platform for isoprene production extremely challenging.

More efforts on engineering *E. coli* for isoprene production often underwent by manipulation of the native DXP/MEP pathway. However, the metabolic flux in this endogenous pathway is subject to strict intracellular regulatory control. Therefore, to rebuild an artemisinin biosynthetic pathway in *E. coli*, Martin et al. (2003) bypassed the DXP/MEP pathway by introducing entire MVA pathway genes from yeast. These genes were organized into two artificial operons, in which *MevT* operon is responsible for conversion from acetyl-CoA to mevalonate, and *MBIS* operon is involving generation of FPP from mevalonate. Re-construction of the MVA pathway in *E. coli*, however, resulted in severe growth inhibition due to the excessive accumulation of FPP, IPP and/or DMAPP. Co-expression of a codon optimized *ADS* gene alleviated the intermediate toxicity and finally yielded amorpha-4,11-diene with a titer of 280-480 mg/L in a fed-batch bioreactor (Newman et al. 2006).

While expression of *MevT* operon led to HMG-CoA toxicity in *E. coli*, co-expression of *HMGR* gene prevented the harmful buildup of HMG-CoA and increased mevalonate production for almost twofold (Pitera et al. 2007). Manipulation of the intergenic *MevT* operon regions strongly impacted the balancing expression pattern of individual genes and consequently resulted in sevenfold increase in the mevalonate titer (Pfleger et al. 2006). Enhancement of gene stability by chromosomal integration of the MVA pathway genes also greatly enhanced mevalonate production (Yuan et al. 2006). These efforts demonstrated that the carbon source can be effectively channeled to the exogenous isoprene pathway in the engineered *E. coli* strain even though the flux of native isoprene biosynthesis remains a low level.

The recombinant sesquiterpene synthases appear to lack the specificity that their native equivalents should have. For example, the engineered bacteria-expressed *Abies grandis* δ-selinene synthase and γ-humulene synthase give rise to 34 and 52 sesquiterpenes, respectively (Steele and Crock 1998). Similarly, the recombinant *A. annua* ADS enzyme generates 16 sesquiterpenes. However, the fidelity of recombinant ADS-catalyzed reaction in yeast can be considerably improved in presence of the divalent metal irons, Mn^{2+} or Co^{2+} (Picaud et al. 2005). The byproduct (δ-cadinene) of peppermint (*E*)-β-farnesene synthase can be eliminated as Mn^{2+} presents (Crock et al. 1997). Two recombinant maize sesquiterpene synthases enable production of at least 23 sesquiterpenes, but their product spectra shift toward (*E*)-β-farnesene with Mn^{2+} supplement (Kollner et al. 2004).

Functional expression of the plant-isolated *CYP* genes in *E. coli* represents a challenge in proper domain folding, membrane insertion, cofactor incorporation, post-translational modification and essential factor interactions (Carter et al. 2003). By exploiting some of plant-derived *CYP* genes, Chang et al. (2007) demonstrated that *E. coli* can be genetically engineered to produce high levels of complex sesquiterpenes. Although expression of the native *CYP71AV1* gene only led to the undetectable CYP activity, codon optimization coupled with transmembrane domain modification resulted in generation of some intermediates such as artemisinic alcohol at a relative low concentration (0.18-0.45 mg/L). Replacement of *Candida tropicalis CPR* gene (*ctCPR*) with *A. annua CPR* gene (*aaCPRct*) increased artemisinic alcohol for 12-fold to 5.6 mg/L. Furthermore, choice of the plasmid vectors and *E. coli* strains can affect the artemisinic acid titer. Use of pCWori plasmid and *E. coli* strain DH1 further increased the oxidized product of amorpha-4, 11-diene by more than 1000-fold to 553 mg/L.

7.3. Engineering Microbes as Factories for Total Artemisinin Production

Why expression of *CYP71AV1* gene in engineered microbes only led to generation of artemisinic acid rather than artemisinin? We argue that the catalytic reaction of CYP71AV1 enzyme may be changed in microbial circumstances where it is only able to catalyze conversion from amorpha-4,11-diene to artemisinic acid; or CYP71AV1 enzyme is actually another type of cytochrome P450 enzyme (CYP) that naturally produces artemisinic acid. In other words, the genuine CYP gene *bona fide* responsible for generation of dihydroartemisinic acid may not be isolated until now. In fact, *CYP71AV1* gene that was first cloned by the Covello group (Teoh et al. 2006) and by the Keasling group (Ro et al. 2006) is both from amplification of the *A. annua* CYP by a pair of degenerate primers designed according to the bioinformatic analysis data. Therefore, Covello et al. (2007) have suggested that other genes encoding the putative double bond reductase and possibly corresponding enzymes presumably allow bioengineering toward dihydroartemisinic acid production (and possibly artemisinin) and avoid the chemical steps required to convert artemisinic acid to dihydroartemisinic acid.

There have two alternative options for us: the first one is pursuing to find out the original CYP gene responsible for dihydroartemisinic acid biosynthesis; the second one is to modify the currently isolated CYP gene to evolved it into a novel CYP gene that encodes an enzyme mimicking the artemisinin biosynthetic reaction. For the former strategy, one can browse the CYP gene website (http://drnelson.utmem.edu/CytochromeP450.html) and check out the candidate CYP genes among more than 1000 accessed CYP genes to help designing hybridization probes for further cDNA library screening. CYP represents a huge and complicated gene family whose members show a wide range of homology from 1-100%, so it must be a great challenge for the sequence-dependent CYP gene isolation; For the later strategy, it will be promise to use the presently isolated CYP genes as the start templates for their artificial evolution by the error-prone PCR or DNA shuffling procesure.

The site-directed artificial evolution of enzymes is the most efficient strategy that well simulate the natural evolution, because the artificial evolution can be promptly accomplished *in vitro*, thereby enabling the evolutionary process shortening from more than million years to a couple of years or even several months (Stemmer 1994). For example, the stereo-specificity of a kind of S-transaminase has increased from 65% to 80-94% only screening approximately 10 000 random mutants (Matchem and Bowen 1996). It may be a preferential choice to employ DNA shuffling method among the homologous sequences for CYP gene mutagenesis, because the benefit mutation can be quickly accumulated within the desired segments.

CONCLUSION

Terpenes are synthesized by the complicated reactions associated with two independent anabolistic pathways that operate in plants, i.e. the MVA pathway in the cytosol and the MEP/DXP pathway in plastids. Their interactions provide an array of terpenes that regulate plant growth and development, and mediate plant-environment interactions. Intensive attempts to manipulate terpene anabolism have been carried out in many plant species for generating valuable metabolites that may meet the industrial and commercial needs, for evaluating contribution of the specific classes of terpenes to the life spans, and for annotating the function of putative terpene biosynthetic genes.

Engineering of terpene metabolism in plants is an attractive goal due to their elaborate biosynthetic potential and obvious economic benefit conferred by the photosynthesis- driven energy and biomass production system. The appreciable successes by distinct strategies dealing with genetic modification of *A. annua*, such as the upstream pathway manipulation, antisense-based modification and compartment protein targeting, etc. have been deeply impressed although much spaces for improvement and optimization left yet.

Overexploitation of the natural products as the high-valued fine chemicals has stirred interests in establishing the alternative production platform to facilitate preservation of the endangered species. Recent efforts have succeeded to engineer *E. coli* with a combination of the MVA pathway genes and the MEP/DXP pathway genes. At the same time, *S. cerevisiae* has also been developed as an artemisinin production platform by which a high-level artemisinic acid was synthesized upon overexpression of *A. annua* genes.

Biosynthesis and accumulation of artemisinin is genetically and environmentally dependent in *A. annua* and can be modulated through two ways of interventions-genetic and environmental. From the genetic view, not only the chemical types with highly concentration of secondary messengers such as JA and MJ are presented in *Artemisia*, but also those with significantly different content of artemisinin and its precursors have been found in *A. annua* although it is unclear whether the former is one of the causes leading to the latter. From the environmental view, because generation of the secondary signals is obviously the reliable outcome of environmental induction, we can maximum the effects of environmental factors on terpene accumulation in the future endeavors toward artemisinin overproduction. In theory, a new *A. annua* plant lines with artemisinin overproduction should be readily created through the metabolic engineering. We may breed the transgenic plants, for example, capable of smoothly redirecting the carbon flux from the plastidic terpenoids or cytosolic steroids to artemisinin production in combination with the environmental induction. Otherwise, artemisinin is certainly enhanced only in a limited range as posed by environmental stimuli.

ACKNOWLEDGMENT

We thank Lu-Lu Yin, Xiao-Ling Xu, Xiao-Mei Zeng, Wen-Jie Lu, Li-Xiang Zeng, Xiao-Xia Guo, Ping-Zu Zhang for their helpful assistance and constructive discussion. This work is partially supported by the Natural Science Foundation of China (NSFC) with the Project Nos. 30672614 to Li-Ling Feng and 30271591 to Qing-Ping Zeng and by Guangdong Provincial Scientific Development Project of China (GPSDPC) under the approved Project No. 2007B031404008 to Qing-Ping Zeng.

REFERENCES

[1] Abdin, M. Z., Israr, M., Srivastava, P. S. & Jain, S. K. (2000). In vitro production of artemisinin, a novel antimalarial compound from *Artemisia annua* L. *J. Med. Arom. Plant Sci.*, *22-4a*, 378-384.

[2] Acton, N. & Klayman, D. L. (1985). Artemisitene, a new sesquiterpene lactone endoperoxide from *Artemisia annua*. *Plant Med.*, *5*, 441–442.

[3] Acton, N., Klayman, D. L. & Rollmann, I. J. (1985). Reductive electrochemical HPLC assay for artemisinin (qinghaosu). *Plant Med.*, *5*, 445–446.

[4] Adam, K. P. & Zapp, J. (1998). Biosynthesis of the isoprene units of chamomile sesquiterpenes. *Phytochemistry*, *48*, 953-959.

[5] Afitlhile, M., Fukushige, H., McCraken, C. & Hildebrand, D. (2005). Allen oxide synthase and hydroperoxide lyase product accumulation in *Artemisia species*. *Plant Sci.*, *169*, 139-146.

[6] Akhila, A., Thakur, R. S. & Popli, S. P. (1987). Biosynthesis of artemisinin in *Artemisia annua*. *Phytochemistry*, *16*, 1927–1930.

[7] Akhila, A., Kumkum, R. & Thakur, R. S. (1990). Biosynthesis of artemisinic acid in *Artemisia annua*. *Phytochemistry, 29*, 2129–2132.

[8] Arano, H. (1964). Cytotaxonomic studies in subfamily Carduoideae of Japanese Compositae.XI. The karyotype analysis in some species of *Artemisia*. *Kromosomo,* 57-59, 1883-1888.

[9] Bagchi, G. D., Jain, D. C. & Kumar, S. (1997). Arteether: a potent plant growth inhibitor from. *Artemisia annua*. *Phytochemistry*, 45, 1131-1134.

[10] Baldi, A. & Dixit, V.K. (2007). Yield enhancement strategies for artemisinin production by suspension cultures of *Artemisia annua*. *Bioresour Technol* doi: 10.1016/j.biotech.2007.06.061.

[11] Beekman, A. C., Barentsen, A. R. W., Woerdenbag, H. J., Van Uden, W., Pras, N., El-Feraly, F. S. & Galal, A. M. (1997). Stereochemistry-dependent cytotoxicity of some artemisinin derivatives. *J. Nat. Prod.*, *60*, 325-327.

[12] Beekman, A. C., Wierenga, P., Woerdenbag, H. J., van Uden, W., Pras, N., Konings, A., El-Feraly, F. S., Galal, A. M. & Wikstrom, H. V. (1998). TI - Artemisinin-derived sesquiterpene lactones as potential antitumour compounds: cytotoxic action against bone marrow and tumour cells. *Plant Med.*, *64*, 615-619.

[13] Bertea, C. M., Freije, J. R., Van der Woude, H., Verstappen, F. W., Perk L, Marquez, V., De Kraker J. W., Posthumus, M. A., Jansen, B. J., De Groot, A., Franssen, M. C. &

Bouwmeester, H. J. (2005). Identification of intermediates and enzymes involved in the early steps of artemisinin biosynthesis in *Artemisia annua. Planta Med.*, 71, 40-47.

[14] Bertea, C. M., Voster, A., Verstappen, F. W., Maffei, M., Beekwilder, J. & Bouwmeester, H. J. (2006). Isoprenoid biosynthesis in *Artemisia annua*: cloning and heterologous expression of a germacrene A synthase from a glandular trichome cDNA library. *Arch Biochem. Biophys*, *448*, 3-12.

[15] Bharel, S., Gulati, A., Abdin, M. Z., Srivastava, P. S., Vishwakarma, R. A., Jain, S. K. (1998). Enzymatic synthesis of artemisinin from natural and synthetic precursors.*J. Nat. Prod., 61*, 633-636.

[16] Bhakuni, R. S., Jain, D. C., Sharma, R. P. & Kumar, S. (2001). Secondary metabolites of *Artemisia annua* and their biological activity. *Curr. Sci.*, *80*, 35-48.

[17] Bick, J. A. & Lange, B. M. (2003). Metabolic cross talk between the cytosolic and plastidic pathways of isoprenoid biosynthesis: unidirectional transport of intermediates the chloroplast envelope membrane. *Arch. Biochem. Biophys*, *415*, 146-154.

[18] Brader, G., Tas, E. & Palva, E. T. (2001). Jusmonate-dependent induction of indole glucosinolates in *Arabidopsis* by culture filtrates of the nonspecific pathogen *Erwinia carotovora. Plant Physiol*, *126*, 849-860.

[19] Bouwmeester, H. J., Wallaart, T. E., Janssen, M. H. A., Van Loo, B., Jansen, B. J., Posthumus, M. A., Schmidt, C. O., De Kraker, J. W., Knig, W. A. & Franssen, M. C. (1999). Amorpha-4, 11-diene synthase catalyses the first probable step in artemisinin biosynthesis. *Phytochemistry*, 52, 843-854.

[20] Brown, G.D. & Sy, L. K. (2004), In vivo transformations of dihydroartmisinic acid in Artemisia annua plants. Tetrahedron, 60,, 1139-1159.

[21] Brown, G. D. Cane, D. E. (1981). Biosynthesis of sesquiterpene. In: *Biosynthesis of isoprenoid compounds.* Porter JW, Spurgeon SL (eds), Vol 1 and 2, John Wiley and Sons, New York, pp. 283-374.

[22] Cane, D. E., Wu, Z., Oliver, J. S. & Hohn, T. M. (1993). Overproduction of soluble trichodiene synthase from Fusarium sporotrichioides in *Escherichia coli. Arch. Biochem. Biophys*, *300*, 416-422.

[23] Chang, M. C. Y. & Keasling, J. D. (2006). Production of isoprenoid pharmaceuticals by engineered microbes. *Nat. Chem. Biol.*, *2*, 674-681.

[24] Chang, M. C. Y., Eachus, R. A., Trieu, W., Ro, D. K. & Keasling, J. D. (2007). Engineering Escherichia coli for production of functionalized terpenoids using plant P450s. *Nat. Chem. Biol.*, *3*, 274-277.

[25] Chang, Y. J., Song, S. H., Park, S. H. & Kim, S. U. (2000). Amorpha-4, 11-diene synthase of *Artemisia annua*: cDNA isolation and bacterial expression of a terpene synthase involved in artemisinin biosynthesis. *Arch. Biochem. Biophys., 383*, 178-184.

[26] Charles, D. J., Simon, J. E., Wood, K. V. & Heinstein, P. (1990). Germplasm variation in artemisinin content of *Artemisia annua* using an alternative method of artemisinin analysis from crude plant extracts. *J. Nat. Prod.*, *53*, 157–160.

[27] Carter, O. A., Peters, R. J. & Croteau, R. (2003). Monoterpen biosynthesis pathway construction in *Escherichia coli. Phytochemistry, 64*, 425-433.

[28] Chen, A. X., Lou, Y. G., Mao, Y. B., Lu, S., Wang, L. J. & Chen, X. Y. (2007). Plant terpenoids: biosynthesis and ecological functions. *J. Integr. Plant Biol.*, *49*, 179-186.

[29] Chen, C. & Chen, Z. (2002). Potentiation of developmentally regulated plant defense response by AtWRKY18, a pathogen-induced Arabidopsis transcription factor. *Plant Physiol.*, *129*, 706-716.

[30] Chen, F. T. & Zhang, G. H. (1987). Studies on several physiological factors on artemisinin synthesis in *Artemisia annua* L. *Plant Physiol, 5*, 26–30.

[31] Chen, P. K., Leather, G. R. & Klayman, D. L. (1987). Allelopathic effect of artemisinin and its related compounds from *Artemisia annua*. *Plant Physiol, 68*, 406.

[32] Connolly, J. D. & Hill, R. A. (1991). *Dictionary of terpenoids*. Vol 1, Mono- and Sesquiterpenoids. Chapman and Hall, London.

[33] Covello, P. S., Teoh, K. H., Polichuk, D. R., Reed, D. W. & Nowak, G. (2007). Functional genomics and the biosynthesis of artemisinin. *Phytochemistry, 68*, 1864-1871.

[34] Crock, J., Wildung, M. & Croteau, R. (1997). Isolation and bacterial expression of a sesquiterpene synthase cDNA clone from peppermint (*Mentha piperita* L.) that produces the aphid alarm pheromone (E)-beta-farnesene. *Proc. Natl. Acad. Sci. USA*, *94,* 12833-12838.

[35] Croteau, R., Kutchan, T. M. & Lewis, N. G. (2000). Natural products (secondary metabolites). In: Buchanan B, Gruissem W, Jones R (eds) *Biochemistry and molecular biology of plants.* American Society of Plant Physiologists. Rockville MD, pp1250-1318.

[36] Cubukcu, B., Bray, D. H., Warhust, D. C., Mericli, A. H., Ozhalay, N. & Sariyar, G. (1990). *In vitro* antimalarial activity of crude extracts and compounds from *Artemisia abrotanum* L. *Phytother Res.*, *4*, 203-204.

[37] Davidovich-Rikanati, R., Sitrit, Y., Tadmor, Y., Iijima, Y., Bilenko, N., Bar, E., Carmona, B., Fallik, E., Dudai, N., Simon, J.E., Pichersky, E. & Lewinsohn, E. (2007). Enrichment of tomato flavor by diversion of the early plastidic terpenoid pathway. *Nat. Biotechnol*, *25*, 899-901.

[38] DeJong, JM., et al. (2006). Genetic engineering of taxol biosynthetic genes in *Saccharomyces cerevisiae*. *Biotechnol. Bioeng*, 93, 212-224.

[39] Delledonne, M., Xia, Y., Dixon, R. A. & Lamb, C. (1998). Nitric oxide functions as a signal in plant disease resistance. *Nature*, *394*, 585-588.

[40] Dhingra, V., Rajoli, C. & Narasu, M. L. (2000). Partial purification of proteins involved in the bioconversion of arteannuin B to artemisinin. *Bioresour Technol*, *73*, 279-282.

[41] Dhingra, V. & Narasu, M. L. (2001). Purification and characterization of an enzyme involved in biochemical transformation of arteannuin B to artemisinin from *Artemisia annua*. *Biochem. Biophys. Res. Commun.*, *281*, 558-561.

[42] Dudareva, N., Andersson, S., Orlova, I., Gatto, N., Reichelt, M., Rhodes, D., Boland, W. & Gershenzon, J. (2005). The nonmevalonate pathway supports both monoterpene and sesquiterpene formation in snapdragon flowers. *Proc. Natl. Acad. Sci. USA*, *102*, 933-938.

[43] Duke, S. O., Vaughn, K. C., Croom, E. M. & ElSohly, H. N. (1987). Artemisinin, a constituent of annual wormwood (*Artemisia annua)*, is a selective phytotoxin. *Weed Sci.*, 35, 499–505.

[44] Duke, S. O. & Paul, R. N. (1993). Development and fine structure of the glandular trichomes of *Artemisia annua* L. *Int. J. Plant Sci.*, *154*, 107–118.

[45] Duke, M. V., Paul, R. N., ElSohly, H. N., Sturtz, G., & Duke, S. O. (1994). Localization of artemisinin and artemisitene in foliar tissues of glanded and glandless biotypes of *Artemisia annua* L. *Int. J. Plant Sci.*, *155*, 365–372.

[46] ElFeraly, F. S., AlMeshal, I. A., AlYahya, M. A. & Hifnawy, M. S. (1986). On the possible role of qinghao acid in the biosynthesis of artemisinin. *Phytochemistry*, *25*, 2777–2778.

[47] ElFeraly, F. S., AlMeshal, I. A., & Khalifa, S. I. (1989). *Epi*-deoxyarteannuin B and 6,7-dehydroartemisinic acid from *Artemisia annua*. *J. Nat. Prod.*, *52*, 196–198.

[48] ElSohly, H. N., Croom, Jr E. M., ElFeraly, F. S. & ElSherei, M. M. (1990). A large-scale extraction technique of artemisinin from *Artemisia annua*. *J. Nat. Prod.*, *53*, 1560–1564.

[49] Ferreira, J. F. S., Simon J. E. & Janick, J. (1995a). Developmental studies of *Artemisia annua*: Flowering and artemisinin production under greenhouse and field conditions. *Plant Med.*, *61*, 167-170.

[50] Ferreira, J. F. S., Simon, J. E. & Janick, J. (1995b). Relationship of artemisinin content of tissue cultured, greenhouse grown, and field grown plants of *Artemisia annua*. *Plant Med.*, *61*, 351–355.

[51] Ferreira, J. F. S. (2007). Nutrient deficiency in the production of artemisinin, dihydroartemisinic acid, and artemisinic acid in *Artemisia annua* L. *J. Agr. Food Chem.*, *55*, 1686-1694.

[52] Farmer, E. E. & Ryan, C. A. (1990). Interplant communication: airborne methyl jasmonate induces synthesis of protein inhibitors in plant leaves. *Proc. Natl. Acad. Sci. USA*, *87*, 7713-7716.

[53] Flugge, U. I. & Gao, W. (2005). Transport of isoprenoid intermediates across chloroplast envelope membranes. *Plant Biol.*, (Stuttg) *7*, 91-97.

[54] Fu, T. M. (1991). Study on five species of *Artemisia* plants in inner Mongolia. *Univ J Inner Mongolia*, (Nat Sci Ed), *22*, 422-427.

[55] Fulzele, D. P., Sipahimalani, A. T. & Heble, M. R. (1995). Tissue culture of *Artemisia annua* L. plant cultures in bioreactor. *J. Biotechnol*, *40*, 139-143.

[56] Geng, S., Ma, M., Ye, H. C., Liu, B. Y., Li, G. F. & Chong, K. (2001). Effects of *ipt* gene expression on the physiological and chemical characteristics of *Artemisia annua L. Plant Sci.*, *160*, 691-698.

[57] Gupta, S. K., Singh, P., Bajpai, P., Ram, G., Singh, D., Gupta, M. M. & Jain, D. C. (2002). Morphogenetic variation for artemisinin and volatile oil in *Artemisia annua*. *India Crop Prod.*, *16*, 217-224.

[58] Helliwell, C. A., Poole, A., Peacock, W. J. & Dennis, E. S. (1999). *Arabidopsis ent*-kaurene oxidase catalyzes three steps of gibberellin biosynthesis. *Plant Physiol*, *119*, 507-510.

[59] Helliwell, C. A., Chandler, P. M., Poole, A., Dennis, E. S. & Peacock, W. J. (2001). The CYP88A cytochrome P450, *ent*-kaurenoic acid oxidase, catalyzes three steps of the gibberellin biosynthesis pathway. *Proc. Natl. Acad. Sci. USA*, *98*, 2065-2070.

[60] Hemmerlin, A., Hoeffler, J. F., Meyer, O., Tritsch, D., Kagan, I. A., Grosdemange-Billiard, C., Rohmer, M. & Bach, T. J. (2003). Cross-talk between the cytosolic mevalonate and the plastidic methylerythritol phosphate pathways in Tobacco bright yellow-2 cells. *J. Biol. Chem.*, *278*, 26666-26676.

[61] Hethelyi, E. B., Cseko, I. B., Grosy, M., Mark, G. & Palinkas, J. J. (1995). Chemical composition of the *Artemisia annua* essential oils from Hungary. *J. Essential Oil Res.*, *7*, 45-48.

[62] Hu, X. Y., Neill, S. J., Cai, W. M. & Tang Z. C. (2003). Nitric oxide mediates elicitor-induced saponin synthesis in cell cultures of *Panax ginseng*. *Funct Plant Biol.*, *30*, 901-907.

[63] Huang, L. & Liu, J. F. (1993). Studies on anti-inflammatory effects of *Artemisia annua*. *China J Chin Mat Med.*, *18*, 44-48.

[64] Irfan, Q. M., Israr, M., Abdin, M. Z. & Iqbal, M. (2005). Response of *Artemisia annua* L. to lead and salt-induced oxidative stress. *Environ. Exp. Bot.*, *53,* 185-193.

[65] Jackson, B. E., Hart-Wells, E. A., Matsuda, S. P. T. (2003). Metabolic engineering to produce sesquiterpenes in yeast. *Org. Lett.*, *5*, 1629-1632.

[66] Jung, M. (1997). Synthesis and cytotoxicity of novel artemisinin analogs. *Bioorg. Med. Chem. Lett.*, *7*, 1091-1094.

[67] Kaplan, B., Davydov, O., Knight, H., Galon, Y., Knight, M.R., Fluhr, R. & Fromm, H. (2006). Rapid transcriptome changes induced by cytosolic Ca^{2+} transients reveal ABRE-related sequences as Ca^{2+}-responsive cis elements in Arabidopsis. *Plant Cell*, *18*, 2733-2748.

[68] Kapoor, R., Chaudhary, V. & Bhatnaga, A. K. (2007). Effects of arbuscular mycorrhiza and phosphorus application on artemisinin concentration in *Artemisia annua* L. *Micorrhiza*, *17*, 581-587.

[69] Kappers, I. F., Aharoni, A., Van Herpen, T. W. J. M., Luckerhoff, L. L. P., Dicke, M. & Bouwmeester, H.J. (2005). Genetic engineering of terpenoid metabolism attracts bodyguards to *Arabidopsis*. *Science*, *309*, 2070-2072.

[70] Kasahara, H., Hanada, A., Kuzuyama, T., Takagi, M., Kamiya, Y. & Yamaguchi, S. (2002). Contribution of the mevalonate and methylerythritol phosphate pathways to the biosynthesis of gibberellins in *Arabidopsis*. *J. Biol. Chem.*, *277*, 45188-45194.

[71] Khosla, C. & Keasling, J. D. (2003). Metabolic engineering for drug discovery and development. *Nat. Rev. Drug Discov.*, *2*, 1019-1025.

[72] Kim, N. C. & Kim, S. U. (1992). Biosynthesis of artemisinin from 11, 12-dihydroarteannuic acid. *J. Korean Agric. Chem. Soc. Rev.*, *35*, 106–109.

[73] Klayman, D. L., Lin, A. J., Acton, N., Scovill, J. P., Hoch, J. M., Milhous, W. K., Theodarides, A. D. & Dobek, A. S. (1984). Isolation of artemisinin(Qinghaosu) from *Artemisia annua* growing in the United States. *J. Nat. Prod.*, *47*, 715–717.

[74] Klayman, D. L. (1993). *Artemisia annua*: from weed to respectable antimalarial plant. In: Kinhorn AD, Balandrin MF (eds) *HumanMedicinal Agents from Plants*, pp. 242–255. American Chemical Society Symposium Series. ACS, Washington, DC.

[75] Knox, J. P. & Dodge, A. D. (1985) Singlet oxygen and plants. *Phytochemistry*, *24*, 889-896.

[76] Kobayashi, K., Suzuki, M., Tang, J., Nagata, N., Ohyama, K., Seki, H., Kiuchi, R., Kaneko, Y., Nakazawa, M., Matsui, M., Matsumoto, S., Yoshida, S. & Muranaka, T. (2007). Lovastatin insensitive 1, a novel pentatricopeptide repeat protein, is a potential regulatory factor of isoprenoid biosynthesis in Arabidopsis. *Plant Cell Physiol.*, *48*, 322-331.

[77] Kollner, T. G., Schnee, C., Gershenzon, J., & Degenhardt, J. (2004). The variability of sesquiterpenes emitted from two *Zea mays* cultivars is controlled by allelic variation of

two terpene synthase genes encoding stereoselective multiple product enzymes. *Plant Cell*, *16*, 1115-1131.

[78] Korenromp, E., Miller, J., Nahlen, B., Wardlaw, T. & Young, M. (2005). World Malaria Report 2005. World Health Organization (WHO), Roll Back Malaria Partnership, Geneva, 2005.

[79] Kreitschitz, A. & Vallès, J. (2003). New or rare data on chromosome numbers in several taxa of the genus *Artemisia* (Asteraceae) in Poland. *Folia Geobotanica*, *38*, 333-343.

[80] Kudakasseril, G. J., Lam, L. & Staba, J. (1987). Effect of sterol inhibitors on the incorporation of ^{14}C-isopentanyl phosphate into artemisinin by cell-free system from *Artemisia annua* cultures and plants. *Plant Med.*, *53*, 280-284.

[81] Kuzamanov, B. A., Georgieva, S. B. & Nikolova, V. A. (1986). Chromosome numbers of Bulgarian flowering plants. I. Family Asteraceae. *Fitologija* (Sofia), *31*, 71-74.

[82] Lange, B. M., Rujan, T., Martin, W. & Croteau, R. (2000). Isoprenoid biosynthesis: the evolution of two ancient and distinct pathways across genomes. *Proc. Natl. Acad. Sci. USA*, *97*, 13172-13177.

[83] Laughlin, J. C. (1993). Effect of agronomic practices on plant yield and anti-malarial constituents of *Artemisia annua* L. *Acta Hortic*, *331*, 53-61.

[84] Laughlin, J. C. (1995). The influence of distribution of antimalarial constituents in *Artemisia annua* L. on time and method of harvest. *Acta Hortic*, *390*, 67-73.

[85] Laughlin JC (2002). Post-harvest drying treatment effects on antimalarial constituents of *Artemisia annua* L. *Acta hortic*, *576*, 315-320.

[86] Laule, O., Furholz, A., Chang, H. S., Zhu, T., Wang, X., Heifetz, P. B., Gruissem, W. & Lange, B. M. (2003). Crosstalk between cytosolic and plastidic pathways of isoprenoid biosynthesis in *Arabidopsis thaliana*. *Proc. Natl. Acad. Sci. USA*, *100*, 6866-6871.

[87] Lee, P. C. & Schmidt-Dannert, C. (2002). Metabolic engineering towards biotechnological production of carotenoids in microorganisms. *Appl. Micro Biotechnol*, *60*, 1-11.

[88] Li, Y., Yang, Z.X., Chen, Y. X. & Zhang, X. (1994). Synthesis of (15_14C) labelled artemisinin. *Acta Pharmaceutica Sinica*, *29*, 713–716.

[89] Li, Y., Huang, H. & Wu, Y. L. (2006). Qinghaosu artemisinin – a fantastic antimalarial drug from a traditional Chinese medicine. In: Liang, XT, Fang WS. (Eds.), *Medicinal Chemistry of Bioactive Natural Products*. John Wiley and Sons Inc, pp. 183–256.

[90] Liersh, R., Soicke, H., Stehr, C. & T¨ullner, H. U. (1986). Formation of artemisinin in *Artemisia annua* during one vegetation period. *Plant Med.*, *52*, 387–390.

[91] Lin, J. Y., Chen, T. S. & Chen, C. S. (1994). Jpn-Kokai Tokkyo Koho Jp, Patent No. 06135830.

[92] Lin, S. Z., Zhang, Z. Y., Lin, Y. Z., Zhang, Q. & Guo, H. (2004). The role of calcium and calmodulin in freezing-induced freezing resistance of *Populus tomentosa* cuttings. *Chinese J. Plant Physiol. Mol. Biol.*, *30*, 59-68.

[93] Lin, R. & Lin, Y. R. (eds.). (1991). Flora of the People's Republic of China. Editorial Committee on Flora of the People's Republic of China of the Chinese Academy of Sciences. Beijing, *Science Press*, *Vol. 76, No. 2*.

[94] Lindahl, A. L., Olsson, M. E., Mercke, P., Tollbom, O., Schelin, J., Brodelius, M. & Brodelius, P. E. (2006). Production of the artemisinin precursor amorpha-4, 11-diene by engineered *Saccharomyces cerevisiae*. *Biotechnol Lett.*, *28*, 571-580.

[95] Liu, C. Z., Wang, Y. C., Ouyang, F., Ye, H. C. & Li, G. F. (1997). Production of artemisinin by hairy root cultures of *Artemisia annua* L. *Biotechnol Lett.*, *19*, 927-929.
[96] Liu J. M., Ni, M. Y., Fan, J. F., Tu, Y. Y., Wu, Z. H., Wu, Y. L. & Chou, W. S. (1979). Structure and reaction of arteannuin. *Acta Chim. Sin*, *37*, 129-143.
[97] Lommen, W. J. M., Schenk, E., Bouwmeester, H. J. & Verstappen, F. W. A. (2006). Trichome dynamics and artemisinin accumulation during development and senescence of *Artemisia annua* leaves. *Plant Med.*, *72*, 336-345.
[98] Lommen, W. J. M., Elzinga, S., Verstappen, F. W. A. & Bouwmeester, H. J. (2007). Artemisinin and sesquiterpene precursors in dead and green leaves of *Artemisia annua* L. crop. *Plant Med.*, *73*, 1133-1139.
[99] Lommen, W. J., Schenk, E., Bouwmeester, H. J. & Verstappen, F. W. (2005). Trichome dynamics and artemisinin accumulation during development and senescence of *Artemisia annua* leaves. *Plant Med.*, *72*, 336-345.
[100] Mahmoud, S. S. & Croteau, R. B. (2002). Strategies for transgenic manipulation of monoterpene biosynthesis in plants. *Trends Plant Sci.*, *7*, 366-373.
[101] Martin, V. J. J., Yoshikuni, Y. & Keasling, J. D. (2001). The in vivo synthesis of plant sesquiterpenes by *Escherichia coli. Biotechnol Bioeng*, *75*, 497-503.
[102] Martin, V. J. J., Pitera, D. J., Withers, S. T., Newman, J. D. & Keasling, J. D. (2003). Engineering a mevalonate pathway in *Escherichia coli* for production of terpenoids. *Nat. Biotechnol*, *21*, 796-802.
[103] Martinez, B. C. & Staba, E. J. (1988). The production of artemisinin in *Artemisia annua* L. tissue cultures. *Adv. Cell Cult.*, *6*, 69–87.
[104] Matcham, G. W. & Bowen, A. R. S. (1996). Biocatalysis for chiral intermediates: meeting commercial and technical challenges. *Chem. Today*, *14*, 20-24.
[105] Menke, F. L. H., Parchmann, S., Mueller, M. J., Kijne, J. W. & Memelink, J. (1999). Involvement of the octadecanoid pathway and protein phosphorylation in fungal elicitor-induced expression of terpenoid indole alkaloid biosynthetic genes in *Catharanthus roseus*. *Plant Physiol.*, *119*, 1289-1296.
[106] Mercke, P., Bengtsson, M., Bouwmeester, H. J., Posthumus, M. A. & Brodelius, P. E. (2000). Molecular cloning, expression, and characterization of amorpha-4, 11-diene synthase, a key enzyme of artemisinin biosynthesis in *Artemisia annua* L. *Arch Biochem. Biophys*, *381*, 173-180.
[107] Morales, M. M., Charles, D. J. & Simon, J. E. (1993). Seasonal accumulation of artemisinin in *Artemisia annua* L. *Acta Hortic*, *344*, 416-420.
[108] Mutabingwa, T. K. (2005). Artemisinin-based combination therapies (ACTs): best hope for malaria treatment but inaccessible to the needy! *Acta Trop.*, *95*, 305-315.
[109] Nair, M. S. R. & Basile, D. V. (1999). Bioconversion of arteannuin B to artemisinin. *J. Nat. Prod.*, *56*, 1559-1566.
[110] Newman, J. D. & Chappell, J. (1999). Isoprenoid biosynthesis in plants: carbon partitioning within the cytoplasmic pathway. *Crit. Rev. Biochem. Mol. Biol.*, *34*, 95-106.
[111] Newman, J. D., Marshall, J., Chang, M. C. Y., Nowroozi, F., Paradise, E., Pitera, D., Newman, K. L. & Keasling, J. D. (2006). High-level production of amorpha-4,11-diene in a two phase partitioning bioreactor of metabolically engineered *Escherichia coli. Biotechnol Bioeng*, 95, 684-691.

[112] Newton, P. & White, N. (1999). Malaria: new development in treatment and prevention. *Ann. Rev. Med.*, *50*, 179-192.

[113] Nishiuchi, T., Shinshi, H. & Suzuki, K. (2004). Rapid and transient activation of transcription of the ERF3 gene by wounding in tobacco leaves: possible involvement of NtWRKYs and autorepression. *J. Biol. Chem.*, *279*, 55355-55361.

[114] Nojiri, H., Sugimori, M., Yamane, H., Nishimura, Y., Yamada, A., Shibuya, N., Kodama, O., Murofushi, N. & Omori, T. (1996). Involvement of jasmonic acid in elicitor-induced phytoalexin production in suspension-cultured rice cells. *Plant Physiol.*, *110*, 387-392.

[115] O'Donnell, P. J., Calvert, C., Atzorn, R., Wasternack, C., Leyser, H. M. O. & Bowles, D. J. (1996). Ethylene as a signal mediating the wound response of tomato plants. *Science*, *274*, 1914-1917.

[116] Paniego, N. B., Maligne, A. E. & Giulietti, A. M. (1993). *Artemisia annua*: *in vitro* culture and the production of artemisinin. In: Bajaj YPS (ed) Medicinal and Aromatic Plants. Biotechnology in Agriculture and Forestry Vol 5, pp. 70–78. Springer-Verlag, Springer.

[117] Paniego, N. B. & Giulietti, A. M. (1996). Artemisinin production by *Artemisia annua* L.-transformed organ culture. *Enzyme Microbiol. Technol.*, *18*, 526-530.

[118] Park, H. C., Kim, M. L., Kang, Y. H., Jeon, J. M., Yoo, J. H., Kim, M. C., Park, C. Y., Jeong, J. C., Moon, B. C., Lee, J. H., Yoon, H. W., Lee, S. H., Chung, W. S., Lim, C. O., Lee, S. Y., Hong, J.C. & Cho, M. J. (2004). Pathogen- and NaCl-induced expression of the SCaM-4 promoter is mediated in part by a GT-1 box that interacts with a GT-1-like transcription factor. *Plant Physiol.*, *135*, 2150-2161.

[119] Pfleger, B. F., Pitera, D. J., Smolke, C. D. & Keasling, J. D. (2006). Combinatorial engineering of intergenic regions in operons tunes expression of multiple genes. *Nat. Biotechnol*, *24*, 1027-1032.

[120] Picaud, S., Olofsson, L., Brodelius, M. & Brodelius, P. E. (2005). Expression, purification and characterization of recombinant amorpha-4,11-diene synthase from *Artemisia annua* L. *Arch. Biochem. Biophys*, *436*, 215-226.

[121] Piel, J., Donath, J., Bandemer, K. & Boland, W. (1998). Mevalonate-independent biosynthesis of terpenoid volatiles in plants - induced and constitutive emission of volatiles. *Angewandte Chemie*, *37*, 2478-2481.

[122] Pitera, D. J., Paddon, C., Newman, J. D. & Keasling, J. D. (2007). Rebuilding a balanced heterologous mevalonate pathway for isoprenoid production in *Escherichia coli*. *Metab Eng.*, *9*, 193-207.

[123] Polya, L. (1949). Chromosome numbers of some Hungarian plants. *Acta Hungarica*, *6*, 12137.

[124] Pras, N., Visser, J. F., Batterman, S., Woerdenbag, H. J., Malingr′e, T. M. & Lugt, C. B. (1991). Laboratory selection of *Artemisia annua* L. for high artemisinin yielding types. *Phytochem Anal*, *2*, 80–83.

[125] Putalun, W., Luealon, W., De-Eknamkul, W., Tanaka, H. & Shoyama, Y. (2007). Improvement of artemisinin production by chitosan in hairy root cultures of *Artemisia annua* L. *Biotechnol. Lett.*, *29*, 1143-1146.

[126] Rieping, M. & Schoffl, F. (1992). Synergistic effect of upstream sequences, CCAAT box elements, and HSE sequences for enhanced expression of chimaeric heat shock genes in transgenic tobacco. *Mol. Gen. Genet*, *231*, 226-232.

[127] Ro, D. K., Paradise, E. M., Ouellet, M., Fisher K. J, Newman K. L, Ndungu, J. M, Ho, KA, Eachus, R. A, Ham, T. S, Kirby, J, Chang, M. C, Withers, S. T, Shiba, Y, Sarpong, R, & Keasling, J. D (2006). Production of the antimalarial drug precursor artemisinic acid in engineered yeast. *Nature, 440*, 940-943.

[128] Rodriguez-Concepcion, M., Fores, O., Martinez-Garcia, J. F., Gonzalez, V., Phillips, M. A., Ferrer, A. & Boronat, A. (2004). Distinct light-mediated pathways regulate the biosynthesis and exchange of isoprenoid precursors during Arabidopsis seedling development. *Plant Cell, 16*, 144-156.

[129] Romero, M. R., Efferth, T., Serrano, M. A., Castano, B., Macias, R. I., Briz, O. & Marin, J. J. (2005). Effect of artemisinin/artesunate as inhibitors of hepatitis B virus production in an in vitro replicative system. *Antiviral Research, 68*, 75–83.

[130] Roth, R. J. & Acton, N. A. (1989). The isolation of Sesquiterpenes from *Artemisia annua*. *J. Chem. Educ., 66*, 349.

[131] Roth, R. J. & Acton, N. A. (1989). A simple conversion of artemisinic acid into artemisinin. *J. Nat. Prod., 52*, 1183–1185.

[132] Sangwan, R. S., Agarwal, K., Luthra, R., Thakur, R. S. & Sangwan, N. S. (1993). Biotransformation of arteannuic acid into arteannuin B and artemisinin in *Artemisia annua*. *Phytochemistry, 34*, 1301-1302.

[133] Sauret-Gueto, S., Botella-Pavia, P., Florres-Perez, U., Martinez-Garcia, J. F., San Roman, C., Leon, P., Boronat, A. & Rodriguez-Concepcion, M. (2006). Plastid cues posttranscriptionally regulate the accumulation of key enzymes of the methylerythritol phosphate pathway in Arabidopsis. *Plant Physiol, 141*, 75-84.

[134] Schuler, M. A. & Werck-Reichhart, D. (2003). Functional genomics of P450s. *Ann Rev Plant Biol., 54*, 629-667.

[135] Schoendorf, A., Rithner, C. D., Williams, A. M. & Croteau, R. B. (2001). Molecular cloning of a cytochrome P450 taxane 10β-hydroxylase cDNA from *Taxus* and functional expression in yeast. *Proc. Natl. Acad. Sci. USA*, 98, 1501-1506.

[136] Shukla, A., Abad Farooqi, A. H., Shukla, Y. N. & Sharma, S. (1992). Effect of triacontanol and chlormequat on growth, plant hormones and artemisinin yield in *Artemisia annua* L. *Plant Growth Regul, 11*, 165.

[137] Simpson, S. D., Nakashima, K., Narusaka, Y., Seki, M., Shinozaki, K., Yamaguchi-& Shinozaki, K. (2003). Two different novel cis-acting elements of erdl, a clpA homologous Arabidopsis gene function in induction by dehydration stress and dark-induced senescence. *Plant J., 33*, 259-270.

[138] Singh, A., Vishwakarma, R. A. & Husain, A. (1988). Evaluation of *Artemisia annua* strains for higher artemisinin production. *Plant Med., 54*, 475–476.

[139] Souret, F. F., Weathers, P. J. & Wobbe, K. K. (2002). The mevalonate independent pathway is expressed in transformed roots of *Artemisia annua* and regulated by light and culture age. *In Vitro Cell Dev. Biol. Plant, 38*, 581-588.

[140] Souret, F. F., Kim, Y., Wyslouzil, B. E., Wobbe, K. K., & Weathers, P. J. (2003). Scale-up of *Artemisia annua* L. hairy root cultures produces complex patterns of terpenoid gene expression. *Biotechnol. Bioeng, 83*, 653-667.

[141] Steele, C. L., Crock, J., Bohlmann, J. & Croteau, R. (1998). Sesquiterpene synthases from grand fir (*Abies grandis*): Comparison of constitutive and wound-induced activities, and cDNA isolation, characterization, and bacterial expression of

delta-selinene synthase and gamma-humulene synthase. *J. Biol. Chem., 273*, 2078-2089.

[142] Steliopoulis, P., Wust, M., Adam, K. P. & Mosandl, A. (2002). Biosynthesis of the sesquiterpene germacrene D in *Solidago Canadensis*: ^{13}C and ^{2}H labeling studies. *Phytochemistry, 60*, 13-20.

[143] Stemmer, W. P. C. (1994). Rapid evolution of a protein *in vitro* by DNA shuffling. *Nature, 370,* 389-391.

[144] Stoessl, A., Stothers, J. B. & Ward, E. W. B. (1976). Sesquiterpenoid stress compounds of the *Solanaceae*. *Phytochemistry, 15*, 855-873.

[145] Sun, C., Palmqvist, S., Olsson, H., Boren, M., Ahlandsberg, S. & Jansson, C. (2003). A novel WRKY transcription factor, SUSIBA2, participates in sugar signaling in barley by binding to the sugar-responsive elements of the iso1 promoter. *Plant Cell, 15*, 2076-2092.

[146] Suzuka, O. (1950). Chromosome numbers in the genus *Artemisia*. *Jap. J. Genet., 25*, 17-18.

[147] Svensson, J. T., Crosatti, C., Campoli, C., Bassi, R., Stanca, A.M., Close, T.J. & Cattivelli, L. (2006). Transcriptome analysis of cold acclimation in barley albina and xantha mutants. *Plant Physiol., 141*, 257-270.

[148] Sy, L. K. & Brown, G. D. (2002). The mechanism of the spontaneous autooxidation of dihydro- artemisinic acid. *Tetrahydron, 58*, 897-908.

[149] Tamogami, S., Rakwal, R. & Kodama, O. (1997). Phytoalexin production elicited by exogenously applied jasmonic acid in rice leaves (*Oryza sativa* L.) is under the control of cytokinins and ascorbic acid. *FEBS Lett., 412*, 61-64.

[150] Tavlik, A. F., Bishop, S. J., Ayalp, A. & EL-Feraly, F. S. (1990). Effects of artemisinin, dihydroartemisinin and arteether on immune responses of normal mice. *Int. J. Immunol. Pharmacol, 12*, 385-389.

[151] Teoh, K. H., Polichuk, D. R., Reed, D. W., Nowak, G. & Covello, P. S. (2006). *Artemisia annua* L. (Asteraceae) trichome- specific cDNAs reveal CYP71AV1, a cytochrome P450 with a key role in the biosynthesis of the antimalarial sesquiterpene lactone artemisinin. *FEBS Lett., 580*, 1411-1416.

[152] Terzaghi, W. B., Cashmore, A. R. (1995). Light-regulated transcription. *Annu. Rev. Plant Physiol. Plant Mol. Biol., 46*, 445-474.

[153] Torrell, M. & Vallès, J. (2001). Genome size in 21 *Artemisia* L. species (Asteraceae, Anthemideae): Systematic, evolutionary, and ecological implications. *Genome, 44*, 231–238.

[154] Towler, M. J. & Weathers, P. J. (2007). Evidence of artemisinin production from IPP stemming from both the mevalonate and the nonmevalonate pathways. *Plant Cell Rep., 26*, 2129-2136.

[155] Trigg, P. I. (1990). Qinghaosu (Artemisinin) as an antimalarial drug. *Econ. Med. Plant Res., 3*, 20–55.

[156] Ueda, J. & Kato, J. (1980). Isolation and identification of a senescence-promoting substance from wormwood (*Artemisia absinthium* L.). *Plant Physiol, 66*, 246-249.

[157] Valles, J. (1987). Aportacion al conocimiento citotaxonomico de ocho taxones Ibericos del genero *Artemisia* L. (Asteraceae, Anthemideae). *Anales Jard Bot Madrid, 44*, 79-96.

[158] Vallès, J., Torrell, M., Garnatje, T., Garcia-Jacas, N., Vilatersana, R. & Susanna, A. (2003). The genus *Artemisia* and its allies: phylogeny of the subtribe Artemisiinae (Asteraceae, Anthemideae) based on nucleotide sequences of nuclear ribosomal DNA internal transcribed spacers (ITS). *Plant Biol.,* (Stuttgart), *5*, 274-284.

[159] Volkova, S. A. & Boyko, E. V. (1986). Chromosome numbers in some species of *Artemisia* from the southern part of the Soviet Far East. *Bot. Zurn.*, *71*, 1693.

[160] Vogeli, U. & Chappell, J. (1998). Induction of sesquiterpene cyclase and suppression of squalene synthase activities in plant cell cultures treated with fungal elicitor. *Plant Physiol, 88*, 1291-1296.

[161] Wallaart, T. E., Van Uden, W., Lubberink, H. G. M., Woerdenbag H. J., Pras, N, & Quax, W. J. (1999). Isolation and identification of dihydroartemisinic acid from *Artemisia annua* and its role in the biosynthesis of artemisinin. *J. Nat. Prod.,* 1999, *62*, 430-433.

[162] Wallaart, T. E., Pras, N. & Quax, W. J. (1999). Isolation and identification of dihydroartemisinic acid hydroperoxide from *Artemisia annua*: a novel biosynthetic precursor of artemisinin. *J. Nat. Prod., 62*, 1160-1162.

[163] Wallaart, T. E., Pras, N., Beekman, A. C. & Quax, W. J. (2000). Seasonal variation of artemisinin and its biosynthetic precursors in plants of *Artemisia annua* of different geographical origin: proof for the existence of chemotypes. *Plant Med., 66*, 57-62.

[164] Wallaart, T. E., Bouwmeester, H. J., Hille, J., Poppinga, L. & Maijers, N. C. A. (2001). Amorpha-4, 11-diene synthase: cloning and functional expression of a key enzyme in the biosynthetic pathway of the novel antimalarial drug artemisinin. *Planta*, *212*, 460-465.

[165] Wang, C. W. (1961). The forests of China, with a survey of grassland and desert vegetations. Harvard University Maria Moors Cabot Foundation No 5, pp. 171–187. Harvard University Cambridge, MA.

[166] Wang, E. M., Wang, R., DeParasis, J., Loughrin, J. H., Gan, S. S. & Wagner, G. J. (2001). Suppression of a P450 hydroxylase gene in plant trichome glands enhances natural-product-based aphid resistance. *Nat. Biotechnol*, *19*, 371-374.

[167] Wang, H., Ye, H. C., Li, G. F., Liu, B. Y. & Chong, K. (2000). Effects of fungal elicitors on cell growth and artemisinin accumulation in hairy root cultures of *Artemisia annua*. *Acta Bot. Sin.*, *42*, 905-909.

[168] Wang, H., Ge, L., Ye, H. C., Chong, K., Liu, B. Y. & Li, G. F. (2004). Studies on the effects of *fpf1* gene on *Artemisia annua* flowering time and on the linkage between flowering and artemisinin biosynthesis. *Plant Med.*, *70,* 347-352.

[169] Wang, H., Liu, Y., Chung, K., Liu, B. Y., Ye, H. C., Li, Z. Q., Yan, F. & Li, G. F. (2007). Earlier flowering induced by over-expression of CO gene does not accompany increase of artemisinin biosynthesis in *Artemisia annua*. *Plant Biol.,* (Stuttgart) *9*, 442-446.

[170] Wang, J. W., Zhang, Z. & Tan, R. X. (2001). Stimulation of artemisinin production in *Artemisia annua* hairy roots by the elicitor from the endophytic *Colletotrichum* sp. *Biotechnol Lett.*, *23*, 857-860.

[171] Wang, J. W., Xia, Z. H. & Tan, R. X. (2002). Elicitation on artemisinin biosynthesis in *Artemisia annua* hairy roots by the oligosaccharide extract from the endophytic *Colletotrichum* sp. B501. *Acta Bot. Sin.*, *44*, 1233-1238.

[172] Wang, J. W. & Wu, J. Y. (2004). Involvement of nitric oxide in elicitor-induced defense responses and secondary metabolism of *Taxus chinensis* cells. *Nitric Oxide, 11*, 298-306.

[173] Wang, Y. & Weathers, P. J. (2007). Sugars proportionately affect artemisinin production. *Plant Cell Rep., 26,* 1073-1081.

[174] Wang, Y., Xia, Z. Q., Zhou, F. Y., Wu, Y. L., Huang, J. J. & Wang, Z. Z. (1988). Studies on the biosynthesis of arteannuin III. Arteannuin acid as a key intermediate in the biosynthesis of arteannuin and arteannuin B. *Acta Chim. Sin., 46*, 386-387.

[175] Wang, Y., Xia, Z. Q., Zhou, F. Y., Wu, Y. L., Huang, J. J. & Wang, Z. Z. (1993). Studies on the biosynthesis of arteannuin IV. The biosynthesis of arteannuin and arteanniun B by the leaf homogenate of *Artemisia annua* L. *Chin J. Chem., 11*, 457-463.

[176] Walker, T. S., Bais, H. P. & Vivanco, J. M. (2002). Jasmonic acid induced hypercin production in *Hypericum perforatum* L. (St. John wort). *Phytochemistry, 60*, 289-293.

[177] Watson, L. E., Bates, P. L., Evans, T. M., Unwin, M. M. & Estes, J. R. (2002). Molecular phylogeny of Subtribe Artemisiinae (Asteraceae), including *Artemisia* and its allied and segregate genera. *BMC Evol. Biol., 2*, 17-28.

[178] Weathers, P. J., Cheetham, R. D., Follansbee, E., & Theoharides, K. (1994). Artemisinin production by transformed roots of *Artemisia annua. Biotechnol. Lett. 16*, 1281–1286.

[179] Weathers, P. J., Bunk, G., & McCoy, M. C. (2005). The effect of phytohormones on growth and artemisinin production in *Artemisia annua* hairy roots. *In Vitro Cell Dev. Biol. Plant, 41*, 47-53.

[180] Weathers, P. J., Elkholy, S. & Wobbe, K. K. (2006). Artemisinin: the biosynthetic pathway and its regulation in *Artemisia annua*, a terpenoids-rich species. *In Vitro Cell Dev. Biol. Plant, 42*, 309-317.

[181] WHO (2001). Antimalarial drug combination therapy: report of a WHO *technical consultation.* WHO/CDS/ RBM/2001/35, reiterated in 2003.

[182] WHO (2003). International pharmacopoeia, 3rd ed., Vol. 5. Geneva.

[183] WHO (2005). WHO Model List of Essential Medicines, 14th ed. (Revised March 2005). Geneva.

[184] Woerdenbag, H. J., Lugt, C. B. & Pras, N. (1990). *Artemisia annua* L.: a source of novel antimalarial drugs. Pharmaceutisch Weekblad, *Sci. Ed, 12*, 169-181.

[185] Woerdenbag, H. J., Bos, R., Salomons, M. C., Hendrika, H., Pras, N. & Malingre, T. M. (1993). Volatile constituents of *Artemisia annua* L. *Flavour Fragrance J, 8*, 131-137.

[186] Woerdenbag, H. J., Lfiers, J. F. J., Van Uden, W., et al. (1993). Production of the new antimalarial drug artemisinin in shoot cultures of *Artemisia annua* L. *Plant Cell Tiss Org. Cult., 32*, 247-257.

[187] Woerdenbag, H. J., Pras, N., Nguyen, G. C., Bui, T. B., Bos, R., Van Uden, W., Pham, V. Y., Nguyen, V. B., Batterman, S. & Lugt, C. B. (1994). Artemisin in, related sesquiterpenes, and essential oil in*Artemisia annua* during a vegetation period in Vietnam. *Plant Med., 60*, 272-275.

[188] Wu, S. Q., Schalk, M., Clark, A., Miles, R. B., Coates, R. & Chapel, J. (2006) Redirection of cytosolic or plastidic isoprenoid precursors elevates terpene production in plants. *Nat Biotech, 24*, 1441-1447.

[189] Xie, Z., Zhang, Z. L., Zou, X., Huang, J., Ruas, P., Thompson, D. & Shen, Q. J. (2005). Annotations and functional analyses of the rice WRKY gene superfamily reveal positive and negative regulators of abscisic acid signaling in aleurone cells. *Plant Physiol, 137*, 176-189.

[190] Xiong, L. M., Schumaker, K. S. & Zhu, J. K. (2002). Cell signaling during cold, drought, and salt stress. *Plant Cell*, S165-S183.

[191] Xu, M. J., Dong, J. F. & Zhu, M. Y. (2004). Involvement of NO in fungal elicitor-induced activation of PAL and stimulation of taxol synthesis in *Taxus chinenesis* suspension cells. *Chinese Sci. Bull, 49*, 1038-1043.

[192] Xu, M. J., Dong, J. F. & Zhu, M. Y. (2005). Nitric oxide mediates the fungal elicitor-induced hypericin production of *Hypericum perforatum* cell suspension cultures through a jasmonic-acid-dependent signal pathway. *Plant Physiol, 139*, 991-998.

[193] Xu, M. J, Dong, J. F. & Zhu, M. Y. (2005). Effect of nitric oxide on catharanthine production and growth of *Catharanthus roseus* suspension cells. *Biotechno. Bioeng, 89*, 367-371.

[194] Xu, M. J., Dong, J. F. & Zhu, M. Y. (2006). Nitric oxide mediates the fungal elicitor-induced puerarin biosynthesis in *Pueraria thomsonii* Benth. suspension cells through a salicylic acid (SA)-dependent and a jasmonic acid (JA)-dependent signal pathway. *Sci. China Ser. C, 49*, 379-389.

[195] Xue, G. P. (2002). Characterisation of the DNA-binding profile of barley HvCBF1 using an enzymatic method for rapid, quatitative and high-throughput analysis of the DNA-binding activity. *Nucleic Acids Res., 30*, e77.

[196] Xue, G. P. (2003). The DNA-binding activity of an AP-2 transcriptional activator HvCBF2 involved in regulation of low-temperature responsive genes in barley is modulated by temperature. *Plant J., 33*, 373-383.

[197] Yin, L. L., Zhao, C., Huang, Y., Yang, R. Y. & Zeng, Q. P. (2008). Abiotic stress-induced expression of artemisinin biosynthesis genes in *Artemisia annua* L. *Chin J. Appl. Environ. Biol., 14* , 1-5.

[198] Yu, D., Chen, C. & Chen, Z. (2001). Evidence for an important role of WRKY DNA binding proteins in the regulation of NPR1 gene expression. *Plant Cell, 13*, 1527-1540.

[199] Yuan, L. Z., Rouviere, P. E., Larossa, R. A., & Suh, W. (2006). Chromosomal promoter replacement of the isoprenoid pathway for enhancing carotenoids production in *E.coli. Metab. Eng., 8*, 79-90.

[200] Zhang, Y. S., Ye, H. C., Liu, B. Y., Wang, H. & Li, G. F. (2005). Exogenous GA_3 and flowering induce the conversion of artemisinic acid to artemisinin in *Artemisia annua* plants. *Russ J. Plant Physiol, 52*, 58-62.

[201] Zhang, Z. L., Xie, Z., Zou, X., Casaretto, J., Ho, T. H. & Shen, Q. J. (2004). A rice WRKY gene encodes a transcriptional repressor of the gibberellin signaling pathway in aleurone cells. *Plant Physiol., 134*, 1500-1513.

[202] Zhao, J. & Sakai, K. (2001). Multiple signaling pathways mediate fungal elicitor induced *h*-thujaplicin accumulation in *Cupressus lusitanica* cell cultures. *J. Exp. Bot., 54*, 647-656.

[203] Zeng, Q. P., Qiu, F. & Yuan, L. (2008a). Production of artemisinin by genetically modified microbes. *Biotechnol. Lett., 30*, 581-592.

[204] Zeng, Q. P., Zhao, C., Yin, L. L., Yang, R., Yi., Zeng, X. M., Feng, L. L. & Yang, X. Q. (2008b). Artemisinin biosynthetic cDNA and novel EST cloning and their

quantification on low temperature-induced overexpression. *Sci. China Ser. C, 51,* 232-244.

[205] Zheng, G. Q. (1994). Cytotoxic terpenoids and flavonoids from *Artemisia annua*. *Plant Med.*, *60*, 54-57.

[206] Zheng, L. P., Guo, Y. T., Wang, J. W. & Tan, R. X. (2007). Nitric oxide potentates oligosaccharide- induced artemisinin production in *Artemisia annua* hairy roots. *J Integr Plant Biol.*, *50*, 49–55.

In: Environmental Regulation: Evaluation, Compliance . . . ISBN: 978-1-60741-645-6
Ed: Diederik Meijer and Fillipus De Jong

Chapter 10

REGULATION OF NON-POINT SOURCE POLLUTION AND INCENTIVES FOR GOOD ENVIRONMENTAL PRACTICES - THE CASE OF AGRICULTURE*

***Renan Goetz*[a1], *Yolanda Martinez*[b] *and Angels Xabadia*[a]**
[a]Department of Economics, University of Girona, Spain,
[b]Department of Economic Analysis, University of Zaragoza, Spain

INTRODUCTION

The early stages of environmental policy implementation were characterized by the regulation of point source pollution. This is partly explained by the easy identification of these sources and the broad and strong political support for their regulation. In contrast, the regulation of non-point sources began much later and is still being implemented. For instance, international agreements to reduce water and gaseous emissions within the EU are likely to lead to more regulations in European countries in coming years. The European Water Framework Directive (WFD) has great power to reduce non-point pollution in European member states. This initiative is supported by European Environmental Agency findings (2006) that point to agricultural non-point pollution as the primary cause of water quality deterioration in many European watersheds. From an economic perspective, agricultural non-point control involves difficult planning problems characterized by complexity, uncertainty and policy conflicts. The major feature of non-point pollution is that emissions are either not observable or cannot be observed at a reasonable cost. Therefore it is impossible to attribute emissions to particular polluters, and the use of first-best instruments is infeasible. Unfortunately, the economic literature does not clearly indicate which are the optimal second-best instruments to regulate non-point sources.

---

* A version of this chapter was also published in *Ecological Economics Research Trends*, edited by Carolyn C. Pertsova published by Nova Science Publishers, Inc. It was submitted for appropriate modifications in an effort to encourage wider dissemination of research.

1 Corresponding author: Renan Goetz, University of Girona, Department of Economics, Campus Montilivi, s/n, 17071 Girona, Spain, email:renan.goetz@udg.edu, Tel.: 0034 972 418719, Fax.: 0034 972418032.

Given the difficulty of metering discharges with a reasonable degree of accuracy, Segerson (1988), Xepapadeas (1991) and Xepapadeas (1992) proposed taxing the concentration of the pollutant in the environmental media receiving the emissions (ambient tax). Even though recent work by Horan et al. (1998), Hansen (1998) and Hansen and Romstad (2006) has reduced the informational requirements on the regulator to implement ambient taxes, their political acceptability may still be severely limited as there is currently no direct relationship between individual behaviour and the size of the actual tax (Shortle and Abler, 1998).

Another strand of the economic literature focuses on instruments to control pollution indirectly, for instance, regulating inputs or the management practices of the firm. However, in order to be applicable, regulating inputs or management practices has to be truncated to a subset of choices that are easy to observe and highly correlated with pollutant emissions. The restriction on the number of inputs considered limits these approaches to being second-best. Studies like the ones by Mapp et al. (1994), Larson, Helfand and House (1996) and Vickner et al. (1998) analyze the environmental and economic impact of regulating a contaminating input, either by restrictions on the choice of the inputs or by a change in the applied technology. Among the techniques for the second-best regulation of inputs, little attention has been paid to the choice of management practices. Other policy options like voluntary controls resulting from moral suasion and/or education, or economic incentives for farmers to reduce emissions by changing or modifying their farming practices have also received little attention in the literature.[1]

Designing an environmental policy is very difficult in practice. Due to the lack of powerful instruments, regulation in EU member states is concentrated on control instruments like technology related standards and management rules. But these instruments often do not reveal economic incentive and farmers do not act voluntarily.[2] Consequently the introduction of clean technologies and/or the implementation of good environmental practices requires continuous and strict controls which in turn lead to high costs for the regulator. In addition, since information problems prevent the use of first-best instruments, a theoretical rationale for combining instruments may exist.

In this paper we propose a combination of incentives (deposit-refund system) to encourage the adoption of good environmental practices to reduce nitrate emissions due to livestock management. The main objective of this study is to describe the equilibrium conditions that make adapting good environmental practices desirable from social and private points of view. For this purpose we design a specific tax on a polluting input (deposit) and a subsidy for the voluntary adoption of good environmental practices (refund). In contrast to previous work (Fullerton and Wolverton, 2000), the deposit refund system is not linked to output or input but the way the input is applied. As the correct application of good environmental practices cannot be observed by the regulator, the payment of the refund does not depend on any control exercised directly by the regulator. Instead, the payment depends on the presentation of a certificate, issued by independent persons or firms certified to apply the polluting input. Those persons or firms guarantee with their reputation and future business

---

[2] For an excellent and still not outdated overview of the state of the art in non-point source pollution control see Shortle and Abler (1998) and Shortle et al. (1998).

[3] The certification of management practices may offer economic incentives if it refers to the quality of the produced good, for instance organically produced goods.

perspectives that the polluting input has been applied in accordance with good environmental practices. In practice, the certified person can be a farmer of the region who has obtained a license to apply the polluting input.

The theoretical analysis is presented in the following section and thereafter we present an empirical analysis for the optimal management of livestock and cultivation activities.

ECONOMIC PROBLEM

To be more specific we consider a farmer that fattens hogs and cultivates corn. The farmer uses livestock slurry as a fertilizer for the production of the crop. The total net benefits obtained by the farmer originate from the net benefits of pig farming, denoted by π_p, and from the net benefits of corn production, denoted by π_c.

The net benefits from hog production basically depend on the number of places available for the animals. Hence we have:

$$\pi_p = p_p \cdot p - k_f \tag{1}$$

where p_p are the net benefits from each place (€/place), p is the number of places and k_f is the fixed cost of the installation.

As a consequence of hog production, a certain amount of slurry is generated that the farmer uses as a fertilizer in corn production. The farmer can choose between two different technologies, $j=i,s$, for the application of the slurry: a cheap but highly polluting technology (insecure technology) or a less polluting, more costly technology (secure technology). The subscript $j = i$ represents the insecure technology, and $j = s$ denotes the secure technology. The amount of nitrogen contained in the slurry and applied with technology j is denoted by n_j. The costs of the application of the slurry are denoted by p_j (€/kg N). All prices and costs cannot be influenced by the farmer and are therefore considered as given data. The amount of nitrogen assimilated by the crop is a function of the amount of the slurry applied, and is given by $n_a = \sum_j g_j(n_j)$, with $dg_j / dn_j \geq 0;\ d^2g_j / dn_j^2 < 0$. We assume that the secure technology uses the nitrogen more effectively, thus, for any given level of applied nitrogen $\hat{n},\ g_s(\hat{n}) \geq g_i(\hat{n})$.

The function $y(n_a)$ denotes the crop production function (Tm/ha) and depends on nitrogen taken up by the plant. It has the regular properties of a neoclassical production function, i.e., $y_{n_a} > 0, y_{n_a n_a} < 0$.

Given the preceding definitions, the net benefits per hectare from corn production are given by:

$$\pi_c = \sum_j p_y \cdot y(n_a) - p_j \cdot n_j \cdot \delta_j - c + s \tag{2}$$

where p_y denotes the crop price (€/Tm.ha), c is the total fixed cost for corn production (€/ha), and s denotes the direct payments according to the Common Agricultural Policy (CAP) (€/ha). Finally, δ_j reflects the choice of the technology, namely, δ_j=1 if technology j is employed and δ_j=0 if it is not.

The farmer can choose the technology, the amount of organic fertilizer and the number of places available for hog production. We assume that the farmer's objective is to maximize the total net benefits, and hence the decision problem can be formally stated as:

$$\underset{n_i, n_s, \delta_i, \delta_s, p}{Max} \quad \pi_p + \pi_c \tag{3}$$

subject to

$$n_i\delta_i + n_s\delta_s = u \cdot p \tag{4}$$

$$\delta_i + \delta_s = 1 \tag{5}$$

$$n_j \geq 0, j = i, s, \; 0 \leq p \leq P. \tag{6}$$

Condition (4) requires a balance between the amount of nitrogen applied with the slurry and the amount of nitrogen generated by hog production, where u indicates the amount of nitrogen generated by each place (kgN/place). Condition (5) establishes that both technologies cannot be used simultaneously, and equation (6) presents constraints on the admissible values of the control variables.

Taking account of the constraints leads to the Lagrangian:

$$L = p_p \cdot p - k_f + \sum_j \; p_y \cdot y(g(n_j)) - p_j \cdot n_j \;\; \cdot \delta_j - c + s + \mu_1 \;\; n_i\delta_i + n_s\delta_s - u \cdot p \;\; +$$

$$\mu_2 \;\; \delta_i + \delta_s - 1 \;\; + \mu_3 n_i + \mu_4 n_s + \mu_5 p + \mu_6 (P - p)$$

where $\mu_1, \ldots, \mu_6$ are Lagrangian multipliers.

A solution to the problem has to satisfy the following necessary conditions:

$$\frac{\partial L}{\partial n_i} = \left(p_y \cdot \frac{dy}{dn_i} - p_i + \mu_1 \right) \delta_i + \mu_3 = 0 \tag{7}$$

$$\frac{\partial L}{\partial n_s} = \left(p_y \cdot \frac{dy}{dn_s} - p_s + \mu_1 \right) \delta_s + \mu_4 = 0 \tag{8}$$

$$\frac{\partial L}{\partial \delta_i} = p_y \cdot y(g_i(n_i)) - p_i \cdot n_i + \mu_1 n_i + \mu_2 = 0 \tag{9}$$

$$\frac{\partial L}{\partial \delta_s} = p_y \cdot y(g_s(n_s)) - p_s \cdot n_s + \mu_1 n_s + \mu_2 = 0 \tag{10}$$

$$\frac{\partial L}{\partial p} = p_p - \mu_4 u + \mu_5 - \mu_6 = 0 \tag{11}$$

$$n_i \delta_i + n_s \delta_s = u \cdot p \tag{12}$$

$$\delta_i + \delta_s = 1 \tag{13}$$

Necessary conditions (7) and (8) indicate for an interior solution that nitrogen should be employed for each technology up to the point where the value of the marginal product of applied nitrogen plus the shadow value of hog production equals the marginal cost of nitrogen use. Necessary conditions (9) and (10) determine that the technology with greater benefits should be chosen. Condition (12) establishes that all the generated nitrogen must be employed on the farm. Hence, equation (11) establishes for an interior solution that the net benefits from each place should equal the marginal cost given by the shadow value of hog production multiplied by the amount of nitrogen produced by each place. For an interior solution of equation (11) one expects the term $\mu_4 u$ to be positive, representing the marginal cost of the generated nitrogen from an additional place. However, if $p = P$ it could be the case that there is a shortage of nitrogen on the farm, and $\mu_4 u$ presents the marginal value of the generated nitrogen from an additional place. Equation (12), and the fact that p_p is strictly positive, guarantees that the solution of the first order condition (11) with respect to the number of places is finite.

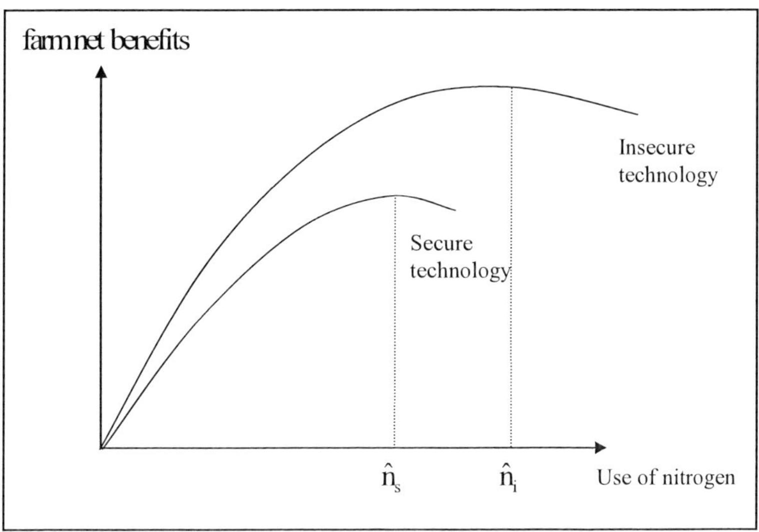

Figure 1. The farm net benfits as a function of nitrogen.

Figure 1 presents the farm net benefits obtained with the insecure and secure technologies. In order to have an interesting economic problem we assume that the secure technology leads to lower farm net benefits than the insecure technology because of its higher associated costs. Consequently, the farmer will adopt the insecure technology since it generates higher profits.

The portion of nitrogen not assimilated by the crop percolates or runs off in the form of diffuse pollution leading to an accumulation of nitrate in aquifers or surface waters. The nitrogen emissions when technology *j* is used are given by $n_j - g_j(n_j)$, and total emissions by:

$$E = \sum_j n_j - g_j(n_j) \cdot \delta_j \tag{14}$$

Since the insecure technology is less efficient with respect to the assimilation of nitrogen by the crop, it causes greater environmental damage than the secure technology. Let MD_j define the marginal monetary damage function of nitrogen emissions for technology *j*.[3]

Moreover, given technology *i*, let $\hat{\pi}$ denote the maximized net benefits of the farmer in the absence of any environmental policy. The optimal input level for the firm for the unconstrained profit maximum is $\hat{n}_i$. Likewise, let π_j denote the constrained maximum net profits of the firm if the level of applied nitrogen, n_j, cannot exceed $\hat{n}_j$, utilizing technology *j*. A reduction in nitrogen leads therefore to a reduction in the farm net benefits. Consequently, the constrained profits will be lower than the unconstrained profits. Let us define the firm's abatement costs C_i and C_s as the difference between unconstrained and constrained profits, that is:

$$C_i = \hat{\pi} - \pi_i$$

$$C_s = \hat{\pi} - \pi_s.$$

The abatement costs are a function of the established upper limit of nitrogen: the lower the limit, the higher the abatement costs of the firm. Given our assumptions about the farm net benefit functions, the marginal abatement costs in the case of the insecure technology are always greater than the marginal abatement costs of the secure technology. Economic theory establishes that the efficient level of emissions is achieved when the marginal abatement costs are equal to the marginal pollution damages. Figure 2 depicts the socially optimal use of nitrogen for each technology.

The farmer maximizes profits by applying $\hat{n}_i$ units of nitrogen. However, since nitrogen leaching causes environmental damage, the farmer's optimum $\hat{n}_i$ and the optimums of the society n_s^* and n_i^* do not coincide. Consequently, farmers will apply nitrogen at socially inefficient rates.

---

[4] For simplicity, the marginal damage functions of Figure 2 are depicted as linear functions, implying that the abatement cost functions are quadratic. Nevertheless other specifications are equally plausible.

If the farmer employing the insecure technology was required to limit the application of nitrogen to n_i^*, the total abatement cost for the farmer would be the area below $MC_i(n)$, between n_i^* and $\hat{n}_i$, that is $\int_{n_i^*}^{\hat{n}_i} MC_i(n)dn$. However, since the avoided monetary damages correspond to the area $\int_{n_i^*}^{\hat{n}_i} MD_i(n)dn$, the overall social gains are positive and correspond to the areas A shown in Figure 2. On the other hand, if farmers employ the secure technology, the optimal level of applied nitrogen is n_s^*. Hence, the total abatement cost for the farmer will be given by the area $\int_{n_s^*}^{\hat{n}_i} MC_i(n)dn$ as a result of the fertilization restriction, and the area $\int_0^{n_s^*} MC_i(n)-MC_s(n)dn$ as a result of the change in the employed technology. In this case, the avoided damages correspond to the area $\int_{n_s^*}^{\hat{n}_i} MD_i(n)dn$ + $\int_0^{n_s^*} MD_i(n)-MD_s(n)dn$. The difference between these areas indicates whether or not it is socially preferred to adopt the secure technology. For instance, area B (avoided damages) in Figure 2 is greater than area C (additional abatement costs) and from a social point of view the secure technology is therefore preferable to the insecure technology. This condition is more likely to hold the lower the difference in application costs and the higher the difference in environmental damages between the two technologies. On the contrary, Figure 3 shows the case where the difference in environmental damages is not significant, and the adaptation of the secure technology is very costly. In this case it is socially optimal to employ the insecure technology.

Given the situation depicted in Figure 2, it would be socially optimal to adopt the secure technology. However, as it reduces their income, farmers are not likely to adopt the secure technology voluntarily.

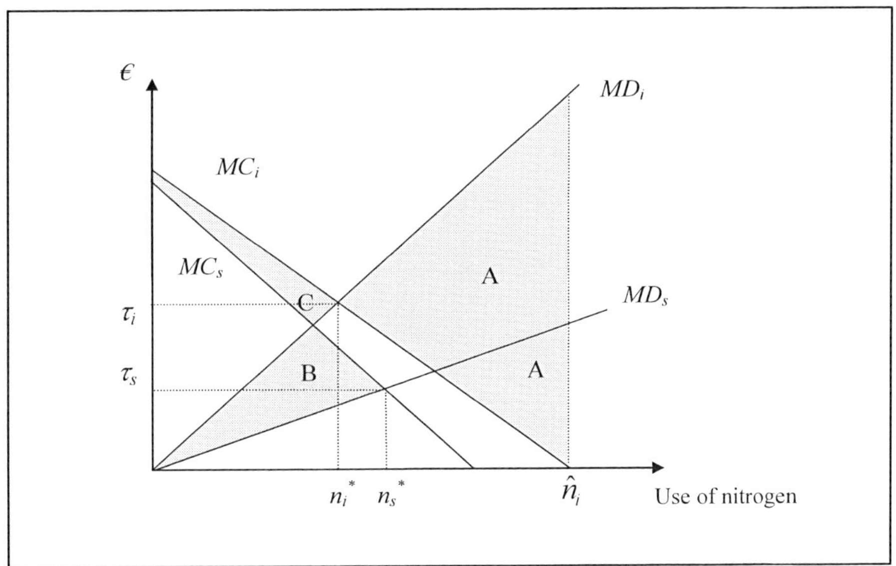

Figure 2. Adoption of the secure technology is socially preferable.

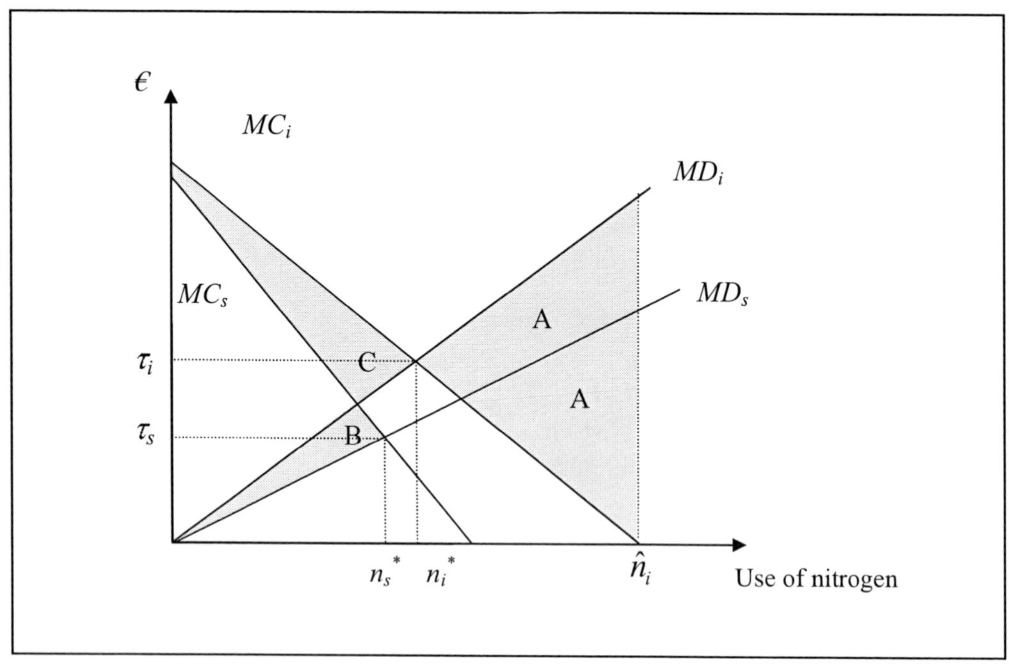

Figure 3. Adoption of the secure technology is socially **not** preferable.

To establish the social optimum in the case where adoption of the secure technology is efficient, we propose a solution based on the combination of a deposit (tax) and a refund (subsidy). In the deposit refund system the voluntary adoption of secure technology maximizes the farmers' net benefits. In the case of hog production the regulator can observe the amount of nitrogen generated at the farm but not whether the farmer employs insecure or secure technology, i.e., bad or good environmental practices. To obtain this information, the regulator creates the figure of a certified person who applies nitrogen in accordance with good environmental practices or oversees its correct application. Once the nitrogen has been applied, the certified person issues a certificate that shows the amount of nitrogen that has been applied correctly. Moreover, the regulator introduces a general tax on nitrogen that has to be paid by all farmers. The optimum level of the nitrogen tax is equal to the marginal cost of abatement at the socially optimal level of nitrogen applied with the insecure technology.

Given technology i and tax τ_i , the total costs for the farmer are the area $\int_{n_i^*}^{\hat{n}_i} MC_i(n)dn$ and the rectangular $\tau_i n_i^*$ (payments of the taxes). These costs correspond to the shaded area, denoted as D, in Figure 4. With the secure technology, abatement costs are $\int_{n_s(\tau_i)}^{\hat{n}_i} MC_i(n)dn$, the payment of the taxes $\tau_i n_s(\tau_i)$, and $\int_0^{n_s(\tau_i)} MC_i(n) - MC_s(n)dn$, that is, the shaded area D plus the area E in Figure 4. Consequently, the farmer will not adopt the secure technology since the abatement costs associated with the secure technology (D+E) are greater than the abatement costs associated with the insecure technology (E). These reflections show that the farmer's decision to adopt the secure technology or not is independent from whether or not the adoption of the secure technology is socially optimal.

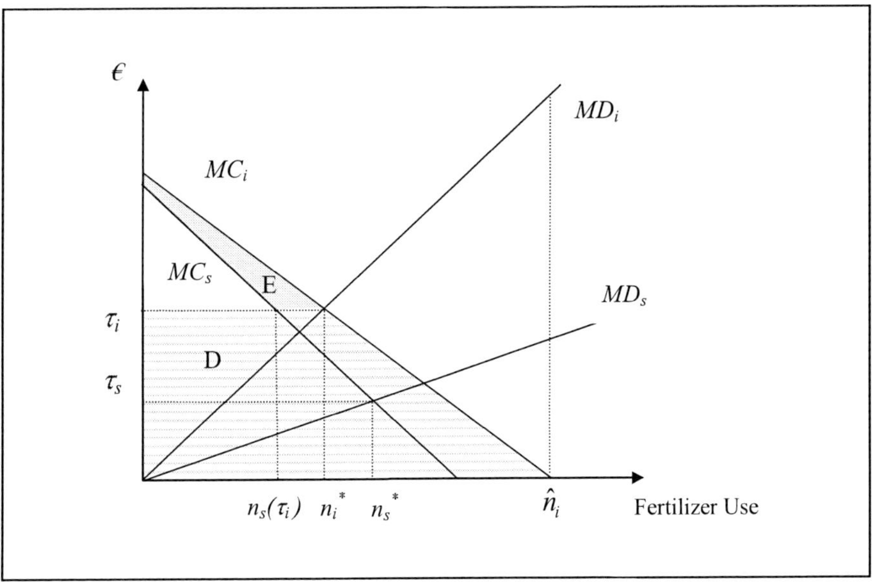

Figure 4. Instruments of a deposit refund system.

To induce the farmer to use the secure technology, the optimum level of the subsidy is given as $\tau_i - \tau_s$. It reflects the decrease in the marginal environmental damage when the secure technology is adopted. If the received subsidy, $(\tau_i - \tau_s)n_s^*$, compensates the additional abatement costs associated with the adoption of the secure technology (area E), the farmer adopts the secure technology and applies the optimal social level of nitrogen n_s^*. In practical terms the farmer will receive refund $\tau_i - \tau_s$ for the amount of nitrogen stated in the certificate obtained from the certified person applying it.

In the following section we illustrate the application of the deposit refund system by using representative data for farms located in the north-eastern part of Spain.

NUMERICAL ILLUSTRATION

In this section we portray the principal agricultural characteristics of the studied area and describe the data employed in the empirical part of this chapter. Furthermore, we specify the parameters and functions of the previously described economic model. Thereafter, we interpret the results of the solution of the model and conduct a sensitivity analysis of the results with respect to values of the marginal environmental damage from nitrogen emissions.

Data and the Area of the Study

Our empirical study is based in Aragon, an autonomous community in the north-eastern part of Spain and one of the main areas of intensive pig production. It accounts for 40% of the total Spanish swine population[4] while representing 54% of the total livestock production and 28% of the final value of the agricultural production (Iguácel 2006). The importance of the pig industry has been growing constantly in Aragon over the last decades, with more than 3.8 million places in 2003, and almost 4 million in 2004 (Iguácel et al. 2005).

For our empirical analysis we considered the operational costs of an average farm located in the study area and representing the behaviour of a farmer with two different yet related production activities: pig and crop production. The specified farm model reproduces the typical conditions of the region with respect to size and biophysical data, with corn being one of the principal crops (see Martínez 2002).

A numerical solution of the mathematical model (equations (3)–(6)) required the functions and parameters to be specified. Relevant data of nitrogen emissions and management costs of different technologies were collected from Iguácel (2006), Daudén et al. (2004) and Daudén and Quílez (2004). Corn production and nitrate leaching were estimated as a function of applied nitrogen, utilising biophysical data, which was previously generated with a process-oriented biophysical model, EPIC (Erosion Productivity Impact Calculator, Mitchell et al. 1998). Production costs, subsidies from the CAP and other data needed to estimate the net benefit functions were determined based on data published annually by the extension service of the Government of Aragon (1999-2007). Additionally, data for the selected region is available from an experimental farm of the Department of Soils and Irrigation (CITA, Government of Aragon)[5]. It includes data about crop production and applied nitrogen. The data generated on this farm allowed us to calibrate our model according to the conditions in the area of the study.[6]

Iguácel and Yagüe (2007) calculated the fixed and variable costs, and determined the efficiency with respect to nitrogen taken up by corn of the insecure and secure technologies. The remaining parameters of the model were determined based on data published annually by the extension service (Government of Aragon 2005, 2007).

Unfortunately, there is no regional data available to estimate water treatment costs as a function of nitrate concentration. Therefore, for simplicity, the function was specified linearly. The unitary cost for water treatment is 1.3 € per kg of nitrate and m^3 of water (Martínez 2002).[7] In Table 1.A of the appendix we present the values of the parameters employed in the empirical study. The identification of the parameters is based on the notation used in the previous section of the chapter. Table 2.A shows the coefficients for the quadratic nitrate leaching functions for each technology j, previously introduced as $n_j - g_j(n_j)$,. The parameters of the functions were estimated using the nonlinear least-squared regression procedure in SHAZAM (White 2002). The economic model was programmed with GAMS (Brooke et al. 1998) and solved numerically using the CONOPT solver.

---

[5] After Germany, Spain has the second largest swine population in the European Union, representing 18% of its total production with a steadily growing trend over the last 10 years (Daudén and Quílez 2004).

[6] See Daudén et al. (2004) for more details with respect to the physical characteristics of the experimental farm.

[7] Details of the conditions of the experiments conducted can be consulted in Daudén et al. (2004) and Daudén and Quílez (2004).

[8] Foess et al. (1998) compared the cost of different processes applied in the USA to remove biological nutrients from water, and reported water treatment costs that range from 1.4 to 21 US$/$m^3$. The great cost discrepancy with respect to our cost can be explained in part because the costs considered here are independent of the pre-treatment nitrate level.

RESULTS AND INTERPRETATION

The application of the economic model requires estimating the farm net benefits, obtained from pig and crop production, as a function of nitrogen, n , (pig slurry) based either on the secure or the insecure technology. The results are presented in Table 1 and show that the farm net benefit functions that best explain the underlying data are quadratic.[8]

Table 1. Benefit Functions with Secure and Insecure Technology

	Insecure technology	Secure technology
Intercept	295.88 (16.23)	70.756 (18.27)
Lineal coefficient (n)	6.0211 (5.31)	6.4396 (7.12)
Square coefficient (n^2)	$-0.79429 * 10^{-2}$ (-8.38)	$-0.79483 * 10^{-2}$ (-10.11)
Adjusted R^2	0.998	0.989

The T Statistics are shown in Parentheses.

After specifying the functions we solved problem (3) – (6) numerically utilizing GAMS. The results are summarized in Table 2. Farm net benefits present the net margin for the individual farmer while private welfare is derived from the farm net benefits by deducting the corresponding taxes and adding up the received subsidies in the case of secure applications. Social welfare before and after the implementation of the policy have to be identical by definition. Hence, the implemented policy achieves its objective of establishing the optimal social outcome.

Table 2. Results with a Marginal Economic Damage of 1.3 € Per Kg of Nitrate Emissions Per M^3

Variables	Insecure technology	Secure technology
Farm net benefit (€/ha)	1432	1374
Nitrogen use from pig slurry (kg/ha)[9]	354	395
Nitrate emissions (kg/ha)	145.24	59.59
Social welfare (€/ha)	1243	1297
Input tax/subsidy (€/kg of nitrogen)	0.39 (tax)	0.23 (subsidy)
Farm net benefits in the presence of a tax/subsidy system (€/ha)	1292	1311
Social welfare in the presence of a tax/subsidy system (€/ha)	1243	1297

[9] We solved a farm decision model with GAMS where the amount of nitrogen that can be applied was restricted. The value of the objective function of this model provided the maximum farm net benefits. Different maximum farm net benefits were obtained by consecutively confining the amount of applied nitrogen from 450 kg/ha to 50 kg/ha. These maximum farm net benefits were then utilized to estimate the farm net benefits as a function of the amount of nitrogen applied.

[10] Water Framework Directive establishes an upper limit for pig slurry application of 170 kgN/ha for vulnerable zones and 250 kgN/ha for the rest of the land. In this paper we have not considered such legal limitation.

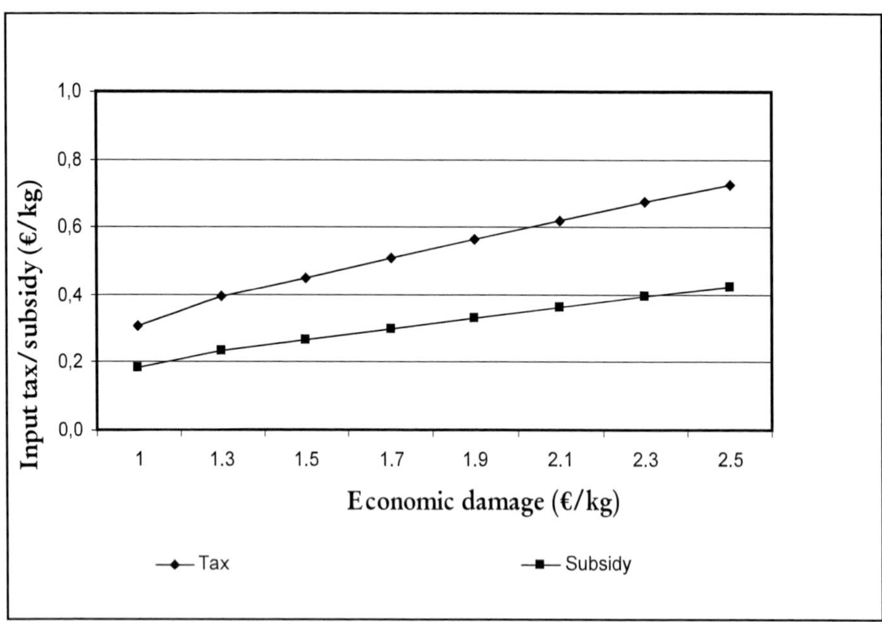

Figure 5. Effect of an increase in the marginal environmental costs of pollution on the tax and subsidy level.

The results of Table 2 show that the farm net benefits with the secure technology are lower than the farm net benefits with the insecure technology. However, the socially optimal amount of applied nitrogen and social welfare are higher with the secure (less polluting) technology. Thus, in the absence of any environmental policy, farmers have no economic incentives to employ the more expensive but less polluting technology that is optimal from a social point of view. To overcome this inefficient private outcome we have designed the deposit/refund system described above so that the adoption of secure technologies is also optimal from a private point of view. Thus, in the presence of the deposit refund system not only does the social welfare have to be higher with secure than with insecure technology, but the private farm net benefits do as well.

In our numerical study, based on a marginal economic damage of 1.3 € per kg of nitrate, a tax of 0.39 €/kg with respect to the applied nitrogen and a subsidy of 0.23 €/kg (see Table 2) induce the socially optimal adoption of secure technology. As the regulator can observe the secure but not the insecure application of nitrogen, all farmers have to pay the tax. However, only the farmers that have adopted the secure technology receive the subsidy. To obtain information about whether individual farmers have adopted the secure technology or not, the regulator relies on the so-called authorized firms. These firms are responsible for the control of secure applications at the farm level.

An increase in water treatment costs from 1 to 2.5 €/kg of nitrate shows that the optimal values of the tax and subsidy have to be changed accordingly. The optimal taxes and subsidies as a response to a change in water treatment costs are presented in Figure 5. As the marginal economic damage of pollution rises, both taxes and subsidies increase. The difference between taxes and subsides increases in absolute terms as the marginal environmental damage increases. However, this difference stays the same in proportional terms since the marginal environmental damage is constant rates and have no incentive to employ good environmental practices. The regulator can observe the total amount of polluting input applied but not whether the farmers follow those good practices or not. To distinguish between farmers who employ good environmental practices and those who do not, the regulator creates the figure of an authorized firm which is responsible for the controlled application of the input. The tax on the polluting input has to be paid by all farmers and the farmers who commission the service of the authorized firm receive a subsidy. This tax/subsidy schemes changes the incentives of farmers so that their privately optimal outcome coincides with the socially optimal outcome.

The proposed solution is illustrated for a farm that is representative of farms located the north-east of Spain. The model calculates the optimal level of the tax and subsidy. It thereby demonstrates how the adoption of good environmental practices can be achieved on a voluntary basis.

APPENDIX

Table 1A. Values of Parameters

Parameters	Values
p_p (€/place)	23.1
k_f (€/place)	18.27
u (m^3/place)	1.57
p_y (€/Tm)	124.83
p_s(€/kg)	2.82
p_i (€/kg)	1.98
c (€/ha)	650.23
s (€/ha)	343.52

Source: Government of Aragon (2005, 2007)

Table 2A. Leaching Function Coefficients.

	Insecure technology	Secure technology
Intercept	75.4 (4.80)	28.42 (5.97)
Lineal coefficient (n)	0.09126 (2.13)	0.034398 (6.27)
Square coefficient (n^2)	$0.299 * 10^{-3}$ (5.39)	$0.1127 * 10^{-3}$ (5.09)
Adjusted R^2	0.89	0.87

The T-Statistics are shown in Parenthesis

ACKNOWLEDGMENTS

Research leading to this paper was supported by the Grants RTA04-141-c2-2 from INIA, AGL2004-00964 from the Spanish Ministry of Education and Science, and 2005SGR00213 from the Department of Innovations, Universities and Firms of the Government of Catalonia.

REFERENCES

Brooke, A., Kendrick, D., Meeraus, A. & Raman, R. (1998). GAMS Tutorial by R. Rosenthal. GAMS Development Corporation. Washington.

Daudén, A. & Quílez, D. (2004). Pig slurry versus mineral fertilization on corn yield and nitrate leaching in a Mediterranean irrigated environment. *European Journal of Agronomy*, 21, pp:7-19.

Daudén, A., Quílez, D. & Vera, M. V. (2004). Pig slurry application and irrigation effects on nitrate leaching in Mediterranean soil lysimeters. *Journal of Environmental Quality*, 33, pp: 2290-2295.

European Environment Agency (2006). EEA Report No 2/2006 Integration of environment into EU agriculture policy—the IRENA indicator-based assessment report, Copenhagen.

Foess, G. W., Steinbrecher, P., Williams, K. & Garret, G.S. (1998). Cost and performance evaluation of BNR processes. *Florida Water Resources Journal December,* 11–16.

Fullerton, D. & Wolverton, A. (2000). Two Generalizations of a Deposit-Refund System, *American Economic Review, Vol. 90, no. 2*, 238-242.

Government of Aragon (1999). Base de datos 1T de superficie de cultivos por término municipal para Aragón 1987-99. Servicio de Estudios y Planificación. Secretaría General Técnica. Departamento de Agricultura. Gobierno de Aragón, Zaragoza.

Government of Aragon (2005). Anuario Estadístico Agrario de Aragón. Departamento de Agricultura y Alimentación.

Government of Aragon (2007). Anuario Estadístico Agrario de Aragón. Departamento de Agricultura y Alimentación.

Iguácel, F., Picot, A. & Gil, M. (2005). Resultados económicos del ganadero de porcino de cebo integrado. Informaciones Técnicas nº 155. Departamento de Agricultura y Alimentación. Gobierno de Aragón.

Iguácel, F. (2006). "Estiércoles y fertilización nitrogenada" en Orús F. coord. Fertilización Nitrogenada, Guía de actualización. Informaciones técnicas, número extraordinario. Departamento de Agricultura y Alimentación. Gobierno de Aragón.

Iguácel, F. & Yagüe, M. R. (2007). Evaluación de costes de sistemas y equipos de aplicación de purín (datos preliminares). Informaciones técnicas nº 178. Departamento de Agricultura y Alimentación. Gobierno de Aragón.

Mapp, H. P., Bernardo, D. J. Sabbagh, G. J., Geleta, S. & Watkins, K. B. (1994). "Economic and Environmental Impacts of Limiting Nitrogen Use to Protect Water Quality: A Stochastic Regional Analysis." *American Journal of Agricultural Economics, 76, no. 4*, November, 889-903.

Martínez, Y. (2002). Análisis económico y ambiental de la contaminación por nitratos en el regadío. Ph.D. Dissertation. University of Zaragoza.

Mitchell, G., Griggs, R., Benson, V. & Williams, J. (1998). The EPIC model: environmental policy integrated climate. Texas Agricultural Experiment Station. Temple.

Larson, D. M., Helfand, G. E. & House, B. W. (1996). "Second-Best Tax Policies to Reduce Nonpoint Source Pollution." *American Journal of Agricultural Economics 78, no. 4* November, 1108-17.

Segerson, K. (1988). "Uncertainty and Incentives for Nonpoint Pollution Control." *Journal of Environmental Economics and Management, 15*, no. *1*, 87-98.

Shortle, J. S. & Abler, D. G. (1998). Nonpoint pollution. *Yearbook of Environmental and Resource Economics,* 1997/1998. Edited by H. Folmer, and T. Tietenberg. Cheltenham: Edwar Elgar.

Shortle, J. S., Horan, R. D. & Abler, D. G. (1998). "Research issues in nonpoint pollution control." *Environmental and Resource Economics, 11(3-4),* 571-585.

Vickner, S. S., Hoag, D. L., Frasier, W. M. & Ascough, J. C. (1998). "A Dynamic Economic Analysis of Nitrate Leaching in Corn Production Under Nonuniform Irrigation Conditions." *American Journal Agricultural Economics, 80, no. 2,* May, 397-408.

White, K. J. (2002). SHAZAM - For Windows, Version 9.0.

Xepapadeas, A. P. (1991). "Environmental Policy under Imperfect Information: Incentives and Moral Hazard." *Journal of Environmental Economics and Management*, 23, no. *2*, 113 - 126.

Xepapadeas, A. P. (1992). "Environmental Policy Design and Dynamic Nonpoint-Source Pollution." *Journal of Environmental Economics and Management, 23, no. 1*, 22-39.

INDEX

A

B

C

D

E

F

G

H

I

J

K

L

M

N

O

P

Q

R

S

T